ENERGY SCIENCE, ENGINEERING AND TECHNOLOGY

CANADIAN OIL SANDS, GREENHOUSE GASES, AND THE KEYSTONE XL PIPELINE

ENERGY SCIENCE, ENGINEERING AND TECHNOLOGY

Additional books in this series can be found on Nova's website under the Series tab.

Additional e-books in this series can be found on Nova's website under the e-book tab.

ENERGY POLICIES, POLITICS AND PRICES

Additional books in this series can be found on Nova's website under the Series tab.

Additional e-books in this series can be found on Nova's website under the e-book tab.

ENERGY SCIENCE, ENGINEERING AND TECHNOLOGY

CANADIAN OIL SANDS, GREENHOUSE GASES, AND THE KEYSTONE XL PIPELINE

TRAVIS J. CARLSON

AND

SHARON HARPER

EDITORS

NOVINKA

New York

For permission to use material from this book please contact us:
Telephone 631-231-7269; Fax 631-231-8175
Web Site: http://www.novapublishers.com

NOTICE TO THE READER

Additional color graphics may be available in the e-book version of this book.

LIBRARY OF CONGRESS CATALOGING-IN-PUBLICATION DATA

ISBN: 978-1-62257-402-5

Published by Nova Science Publishers, Inc. † New York

CONTENTS

PREFACE

After discussing the basic methodology of life-cycle assessments and examining the choice of boundaries, design features, and input assumptions, this book compares several of the publicly available assessments of life-cycle emissions data for Canadian oil sands crudes against each other and against those of other global reference crudes. As congressional concern over the environmental impacts of Canadian oil sands production may encompass both a broad understanding of the global resource as well as a specific assessment of the proposed Keystone XL pipeline, the book surveys both the general scientific literature as well as the individual findings of the State Department's Keystone XL Project Environmental Impact Statement. As lifecycle assessments have become an influential, albeit developing, methodology for collecting, analyzing, and comparing GHG emissions, this book concludes with a discussion of some tools for policymakers who are interested in using these assessments to investigate the potential impacts of U.S. energy policy choices on the environment.

Chapter 1 - *Canadian Oil Sands and Climate Change:* Recent congressional interest in U.S. energy policy has focused in part on ways through which the United States could secure more economical and reliable crude oil resources both domestically and internationally. Many forecasters identify petroleum refined from Canadian oil sands as one possible solution. Increased petroleum production from Canadian oil sands,however, is not without controversy, as many have expressed concern over the potential environmental impacts. These impacts may include increased water and natural gas use, disturbance of mined land, effects on wildlife and water quality, trans-boundary air pollution, and emissions of greenhouse gases (GHG) during extraction and processing. A number of key studies in recent literature have expressed findings that GHG emissions from the production of Canadian oil sands crudes may be higher than those of other crudes imported, refined, and consumed in the United States. The studies identify two main reasons for the increase: (1) oil sands are heavier and more viscous than lighter crude oil types on average, and thus require more energy- and resource-intensive activities to extract; and (2) oil sands are compositionally deficient in hydrogen, and have a higher carbon, sulfur, and heavy metal content than lighter crude oil types on average, and thus require more processing to yield consumable fuels.

Selected Findings from the Primary Published Studies: CRS surveyed the available literature, including the U.S. Department of State-commissioned study in the Environmental Impact Statement for the Keystone XL pipeline project. The literature reveals the following:

- despite differences in the design and input assumptions of the various studies, Canadian oil sands crudes are on average somewhat more GHG emission-intensive than the crudes they would displace in the U.S. refineries, with a range of increase from 14%-20% over the average Well-to-Wheel emissions of other imported crudes;
- discounting the final consumption phase of the life-cycle assessment (which can contribute up to 70%-80% of Well-to-Wheel emissions), Well-to-Tank (i.e., production) emissions from Canadian oil sands crudes have a range of increase from 72%-111% over the average Well-to-Tank emissions of other imported crudes;
- Canadian oil sands crudes, on a Well-to-Wheel basis, range from 9%-19% more emission-intensive than Middle Eastern Sour, 5%-13% more emission-intensive than Mexican Maya, and 2%-18% more emission-intensive than various Venezuelan crudes;
- the estimated effect of the proposed Keystone XL pipeline on the U.S. GHG footprint would be an increase of 3 million to 21 million metric tons of GHG emissions annually (equal to the annual GHG emissions from the combustion of fuels in approximately 588,000 to 4,061,000 passenger vehicles); and
- the estimated effect of the Keystone XL pipeline on global GHG emissions remains uncertain, as some speculate that its construction would encourage an expansion of oil sands development, while others suggest that the project would not substantially influence either the rate or magnitude of oil extraction activities in Canada or the overall volume of crude oil transported to and refined in the United States.

Scope and Purpose of This Report: After discussing the basic methodology of life-cycle assessments and examining the choice of boundaries, design features, and input assumptions, this report compares several of the publicly available assessments of life-cycle emissions data for Canadian oil sands crudes against each other and against those of other global reference crudes. Further, as congressional concern over the environmental impacts of Canadian oil sands production may encompass both a broad understanding of the global resource as well as a specific assessment of the proposed Keystone XL pipeline, the report surveys both the general scientific literature as well as the individual findings of the State Department's Keystone XL Project Environmental Impact Statement. Finally, as life-cycle assessments have become an influential—albeit developing—methodology for collecting, analyzing, and comparing GHG emissions, the report concludes with a discussion of some tools for policymakers who are interested in using these assessments to investigate the potential impacts of U.S. energy policy choices on the environment.

Chapter 2 - When it comes to future reliable oil supplies, Canada's oil sands will likely account for a greater share of U.S. oil imports. Oil sands account for about 46% of Canada's total oil production and oil sands production is increasing as conventional oil production declines. Since 2004, when a substantial portion of Canada's oil sands were deemed economic, Canada, with about 175 billion barrels of proved oil sands reserves, has ranked second behind Saudi Arabia in oil reserves. Canadian crude oil exports were about 1.82 million barrels per day (mbd) in 2006, of which 1.8 mbd or 99% went to the United States. Canadian crude oil accounts for about 18% of U.S. net imports and about 12% of all U.S. crude oil supply.

Oil sands, a mixture of sand, bitumen (a heavy crude that does not flow naturally), and water, can be mined or the oil can be extracted in-situ using thermal recovery techniques. Typically, oil sands contain about 75% inorganic matter, 10% bitumen, 10% silt and clay, and 5% water. Oil sand is sold in two forms: (1) as a raw bitumen that must be blended with a diluent for transport and (2) as a synthetic crude oil (SCO) after being upgraded to constitute a light crude. Bitumen is a thick tar-like substance that must be upgraded by adding hydrogen or removing some of the carbon.

Exploitation of oil sands in Canada began in 1967, after decades of research and development that began in the early 1900s. The Alberta Research Council (ARC), established by the provincial government in 1921, supported early research on separating bitumen from the sand and other materials. Demonstration projects continued through the 1940s and 1950s. The Great Canadian Oil Sands company (GCOS), established by U.S.-based Sunoco, later renamed Suncor, began commercial production in 1967 at 12,000 barrels per day.

The U.S. experience with oil sands has been much different. The U.S. government collaborated with several major oil companies as early as the 1930s to demonstrate mining of and in-situ production from U.S. oil sand deposits. However, a number of obstacles, including the remote and difficult topography, scattered deposits, and lack of water, have resulted in an uneconomic oil resource base. Only modest amounts are being produced in Utah and California. U.S. oil sands would likely require significant R&D and capital investment over many years to be commercially viable. An issue for Congress might be the level of R&D investment in oil sands over the long term.

As oil sands production in Canada is predicted to increase to 2.8 million barrels per day by 2015, environmental issues are a cause for concern. Air quality, land use, and water availability are all impacted. Socio-economic issues such as housing, skilled labor, traffic, and aboriginal concerns may also become a constraint on growth. Additionally, a royalty regime favorable to the industry has recently been modified to increase revenue to the Alberta government. However, despite these issues and potential constraints, investment in Canadian oil sands will likely continue to be an energy supply strategy for the major oil companies.

Chapter 3 - In 2008, Canadian pipeline company TransCanada filed an application with the U.S. Department of State to build the Keystone XL pipeline, which would transport crude oil from the oil sands region of Alberta, Canada, to refineries on the U.S. Gulf Coast. Keystone XL would ultimately have the capacity to transport 830,000 barrels per day, delivering crude oil to the market hub at Cushing, OK, and further to points in Texas. TransCanada plans to build a pipeline spur so that oil from the Bakken formation in Montana and North Dakota can also be carried on Keystone XL.

As a facility connecting the United States with a foreign country, the pipeline requires a Presidential Permit from the State Department. In evaluating such a permit application, the department must determine whether it is in the "national interest," considering the project's potential effects on the environment, economy, energy security, foreign policy, and other factors. Environmental impacts are considered pursuant to the National Environmental Policy Act, and documented by the State Department in an Environmental Impact Statement (EIS). The final EIS was released for the Keystone XL pipeline permit application in August 2011, after which a 90- day public review period began to make the national interest determination. During that time the State Department determined that more information was needed to consider an alternative pipeline route avoiding the environmentally sensitive Sand Hills region of Nebraska, an extensive sand dune formation with highly porous soil and a shallow depth to groundwater recharging the Ogallala aquifer.

The Temporary Payroll Tax Cut Continuation Act of 2011 (P.L. 112-78) required the Secretary of State to approve or deny the project within 60 days. On January 18, 2012, the State Department, with the President's consent, denied the Keystone XL permit, citing insufficient time under this deadline to properly assess the reconfigured project. Subsequently, TransCanada announced that it would proceed with development of the pipeline segment connecting Cushing, OK, to the Gulf Coast as a stand-alone project not requiring a Presidential Permit—a decision supported by the Obama administration. In April 2012, TransCanada submitted to Nebraska proposed pipeline routes avoiding the Sand Hills. Subsequently, on May 4, 2012, TransCanada submitted a new application for a Presidential Permit that includes proposed new routes through Nebraska. With the new permit application, the NEPA compliance process begins anew, although it may draw from relevant existing analysis and documentation prepared for the August 2011 final EIS.

In the wake of the State Department's denial of the Presidential Permit, Congress has debated legislative options addressing the Keystone XL pipeline. The Surface Transportation Extension Act of 2012, Part II (H.R. 4348) and the North American Energy Access Act (H.R. 3548) would transfer the permitting authority for the Keystone XL pipeline project to the Federal Energy Regulatory Commission, requiring FERC to issue a permit within 30 days of enactment. The Keystone For a Secure Tomorrow Act (H.R. 3811), the Grow America Act of 2012 (S. 2199), S. 2041 (a bill to approve the Keystone XL pipeline), the EXPAND Act (H.R. 4301), and the Energizing America through Employment Act (H.R. 4000) would immediately approve the original permit application filed by TransCanada. All seven bills include provisions allowing for later alteration of the pipeline route in Nebraska. S. 2100 and H.R. 4211 would suspend sales of petroleum products from the Strategic Petroleum Reserve until issuance of a Presidential Permit for the Keystone XL project. Changing or eliminating the State Department's role in issuing cross-border infrastructure permits may raise questions about the President's executive authority, however. H.R. 3900 would seek to ensure that crude oil transported by the Keystone XL pipeline, or resulting refined petroleum products, would be sold only into U.S. markets, but this bill could raise issues related to international trade agreements.

In: Canadian Oil Sands, Greenhouse Gases … ISBN: 978-1-62257-402-5
Editors: T. J. Carlson and S. Harper

Chapter 1

CANADIAN OIL SANDS: LIFE-CYCLE ASSESSMENTS OF GREENHOUSE GAS EMISSIONS*

Richard K. Lattanzio

SUMMARY

Canadian Oil Sands and Climate Change

Recent congressional interest in U.S. energy policy has focused in part on ways through which the United States could secure more economical and reliable crude oil resources both domestically and internationally. Many forecasters identify petroleum refined from Canadian oil sands as one possible solution. Increased petroleum production from Canadian oil sands,however, is not without controversy, as many have expressed concern over the potential environmental impacts. These impacts may include increased water and natural gas use, disturbance of mined land, effects on wildlife and water quality, trans-boundary air pollution, and emissions of greenhouse gases (GHG) during extraction and processing. A number of key studies in recent literature have expressed findings that GHG emissions from the production of Canadian oil sands crudes may be higher than those of other crudes imported, refined, and consumed in the United States. The studies

* This is an edited, reformatted and augmented version of Congressional Research Service., Publication No. R42537, dated May 15, 2012.

identify two main reasons for the increase: (1) oil sands are heavier and more viscous than lighter crude oil types on average, and thus require more energy- and resource-intensive activities to extract; and (2) oil sands are compositionally deficient in hydrogen, and have a higher carbon, sulfur, and heavy metal content than lighter crude oil types on average, and thus require more processing to yield consumable fuels.

Selected Findings from the Primary Published Studies

CRS surveyed the available literature, including the U.S. Department of State-commissioned study in the Environmental Impact Statement for the Keystone XL pipeline project. The literature reveals the following:

- despite differences in the design and input assumptions of the various studies, Canadian oil sands crudes are on average somewhat more GHG emission-intensive than the crudes they would displace in the U.S. refineries, with a range of increase from 14%-20% over the average Well-to-Wheel emissions of other imported crudes;
- discounting the final consumption phase of the life-cycle assessment (which can contribute up to 70%-80% of Well-to-Wheel emissions), Well-to-Tank (i.e., production) emissions from Canadian oil sands crudes have a range of increase from 72%-111% over the average Well-to-Tank emissions of other imported crudes;
- Canadian oil sands crudes, on a Well-to-Wheel basis, range from 9%-19% more emission-intensive than Middle Eastern Sour, 5%-13% more emission-intensive than Mexican Maya, and 2%-18% more emission-intensive than various Venezuelan crudes;
- the estimated effect of the proposed Keystone XL pipeline on the U.S. GHG footprint would be an increase of 3 million to 21 million metric tons of GHG emissions annually (equal to the annual GHG emissions from the combustion of fuels in approximately 588,000 to 4,061,000 passenger vehicles); and
- the estimated effect of the Keystone XL pipeline on global GHG emissions remains uncertain, as some speculate that its construction would encourage an expansion of oil sands development, while others suggest that the project would not substantially influence either the rate or magnitude of oil extraction activities in Canada or the overall volume of crude oil transported to and refined in the United States.

Scope and Purpose of This Report

After discussing the basic methodology of life-cycle assessments and examining the choice of boundaries, design features, and input assumptions, this report compares several of the publicly available assessments of life-cycle emissions data for Canadian oil sands crudes against each other and against those of other global reference crudes. Further, as congressional concern over the environmental impacts of Canadian oil sands production may encompass both a broad understanding of the global resource as well as a specific assessment of the proposed Keystone XL pipeline, the report surveys both the general scientific literature as well as the individual findings of the State Department's Keystone XL Project Environmental Impact Statement. Finally, as life-cycle assessments have become an influential—albeit developing—methodology for collecting, analyzing, and comparing GHG emissions, the report concludes with a discussion of some tools for policymakers who are interested in using these assessments to investigate the potential impacts of U.S. energy policy choices on the environment.

INTRODUCTION

Recent congressional interest in U.S. energy policy has focused in part on ways through which the United States could secure more economical and reliable crude oil resources both domestically and internationally. Many forecasters identify petroleum products refined from Canadian oil sands[1] crude as one possible solution. Canadian oil sands account for about 46% of Canada's total oil production, and that number is expected to rise from its current level of 1.2 million barrels per day (mbd) to 2.8 mbd by 2015.[2] Further, the infrastructure to produce, upgrade, refine, and transport the resource from Canadian oil sand reserves to the United States is in place, and additional infrastructure projects—such as the Keystone XL pipeline—have been proposed.[3] Increased oil production from Canadian oil sands, however, is not without controversy, as many have expressed concern over the potential environmental impacts. These impacts may include increased water and natural gas use, disturbance of mined land, effects on wildlife and water quality, trans-boundary air pollution, and emissions of greenhouse gases (GHG) during extraction and processing.

A number of key studies in recent literature have expressed findings that GHG emissions from the production of Canadian oil sands crudes may be higher than those of other crudes imported, refined, and consumed in the United States.[4] While GHG emissions and other air quality issues originating in the upstream sectors of Canada's petroleum industry do not directly impact U.S. National Emissions Inventories or U.S. GHG reporting per se, many environmental stakeholders and policymakers have noted that the increased use of more emission-intensive resources in the United States may have negative consequences for both U.S. and global energy policy and environmental compliance.

The U.S. Department of State (DOS), in response to comments on the draft Environmental Impact Statement (EIS) for the Keystone XL pipeline project (which would connect oil sands production facilities in the Western Canadian Sedimentary Basin with refinery facilities in the United States), commissioned a contractor's study on the GHG life-cycle emissions associated with these resources in comparison to other reference crudes.[5] DOS presented this analysis in the Final EIS as a "matter of policy," and noted that neither the National Environmental Policy Act (NEPA) nor DOS regulations (22 C.F.R. 161.12) nor Executive Orders 13337 and 12114 (Environmental Effects Abroad of Major Federal Activities) legally require that an EIS include an assessment of environmental activities outside the United States. In the Final EIS, DOS supported the claim that while the proposed Keystone XL pipeline project may contribute to certain trans-boundary and continental scale environmental impacts, it may not substantially influence either the rate or magnitude of oil extraction activities in Canada or the overall volume of crude oil transported to and refined in the United States.[6]

This report presents a summary of a range of life-cycle emissions assessments of Canadian oil sands crudes and provides an analysis of their respective findings. The first section of the report, "Life-Cycle Assessment Methodology," discusses the basic methodology of life-cycle assessments and examines the choice of boundaries, design features, and input assumptions. The second section of the report, "Results of Selected Life-Cycle Emissions Assessments," compares several of the publicly available assessments of life-cycle GHG emissions data for Canadian oil sands crudes against each other, against those of other global reference crudes, and against those of other fossil fuel resources. The third section, "U.S. Carbon Footprint for the Keystone XL Pipeline," examines some of the specific findings of the Department of State's commissioned study for the Keystone XL pipeline. The report concludes with a discussion of some tools for policymakers who are interested in using these assessments to investigate the potential impacts of U.S. energy policy choices on the environment.

LIFE-CYCLE ASSESSMENT METHODOLOGY

Life-cycle assessment (LCA) is an analytic method used for evaluating and comparing the environmental impacts of various products (in this case, the climate change implications of hydrocarbon resources). LCAs can be used in this way to identify, quantify, and track emissions of carbon dioxide and other GHG emissions arising from the development of these hydrocarbon resources, and to express them in a single, universal metric of carbon dioxide equivalent (CO_2e) GHG emissions per unit of fuel or fuel use.[7] The results of an LCA can be used to evaluate the GHG emissions intensity of various stages of the fuel's life cycle, as well as to compare the emissions intensity of one type of fuel or method of production to another.

GHG emissions profiles modeled by most LCAs are based on a set of boundaries commonly referred to as "cradle-to-grave," or, in the case of transportation fuels such as petroleum, "Wellto-Wheel" (WTW). WTW assessments for petroleum-based transportation fuels focus on the emissions associated with the entire life cycle of the fuel, from extraction, transport, and refining of crude oil, to the distribution of refined product (e.g., gasoline, diesel, jet fuel) to retail markets, to the combustion of the fuel in end-use vehicles. Other LCAs (e.g., Well-to-Tank (WTT) or Wellto-Refinery Gate (WTR)) establish different (i.e., more specific) life-cycle boundaries to evaluate emissions (see Figure 1). Inclusion of the final combustion phase allows for the most complete picture of crude oil's impact on GHG emissions, as this phase can contribute up to 70%-80% of WTW emissions. However, other LCAs can be used to highlight the differences in upstream emissions associated with particular stages as well as experiment with certain boundary assumptions. The choice of boundaries is an important component to any LCA: the more specific a study draws the bounds of its assessment, the more amplified the differences may appear across resources.[8]

Because of the complex life cycle of hydrocarbon fuels and the large number of analytical design features that are needed to model their emissions, LCAs must negotiate many variables and uncertainties in available data. Key factors that influence the results of an LCA include (1) composition of the resource that is modeled, (2) extraction process of the resource that is modeled, (3) design factors chosen for the assessment, and (4) assumptions made in the input data for the assessment.

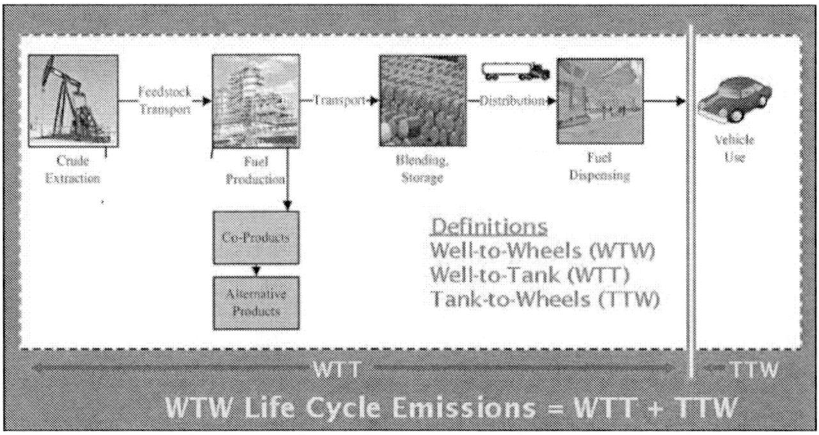

Source: Jacobs Consultancy, *Life Cycle Assessment Comparison of North American and Imported Crudes*, Alberta Energy Research Institute and Jacobs Consultancy, 2009.

Figure 1. Crude Oil Life-Cycle Schematic.

Crude Oil Types. Oil sands are a type of unconventional petroleum deposit. They are commonly loose sand or partially consolidated sandstone containing naturally occurring mixtures of sand, clay, and water, as well as a dense and extremely viscous form of petroleum technically referred to as bitumen.[9] Most LCAs do not include an assessment of raw bitumen, because it is near solid at ambient temperature and cannot be transported in pipelines or processed in conventional refineries. Thus, bitumen is often diluted with liquid hydrocarbons or converted into a synthetic light crude oil to produce the resource known as oil sands crude. Several kinds of crude-like products can be generated from bitumen, and their properties differ in some respects from conventional light crude. They include:

- **Upgraded Bitumen, or Synthetic Crude Oil (SCO).** SCO is produced from bitumen through an upgrading process that turns the very heavy hydrocarbons into lighter fractions. Since the upgrading process begins at the production facility for SCO, the allocation of GHG emissions is weighted more heavily upstream than other crude types.
- **Diluted Bitumen (Dilbit).** Dilbit is bitumen mixed with diluents—typically natural gas liquids such as condensate—to create a lighter, less viscous, and more easily transportable product. Mixing bitumen with less carbon-intensive diluents lessens the GHG emissions impact

per barrel of dilbit in relation to bitumen or SCO. Some refineries need modifications to process large quantities of dilbit feedstock, since it requires more heavy oil conversion capacity than conventional crudes. Increased processing in refineries shifts GHG emissions downstream, potentially intensifying the downstream GHG emission impact of dilbit in relation to SCO or other crudes (e.g., if dilbit is transported from Canada to the United States via a pipeline, the need for increased refining downstream would shift the potential for emissions to the United States).

- **Synthetic Bitumen (Synbit).** Synbit is typically a combination of bitumen and SCO. The properties of each kind of synbit blend vary significantly, but blending the lighter SCO with the heavier bitumen results in a product that more closely resembles conventional crude oil. Refining emissions from synbit occur both upstream and downstream, depending upon a variety of factors.

Extraction Process. Two types of methods for extracting bitumen from the reservoir are currently used in the Canadian oil sands. They include:

- **Mining.** Oil sands deposits that are less than approximately 75 meters below the surface can readily be removed using conventional strip-mining methods. An estimated 20% of currently recoverable reserves are close enough to be mined. The strip-mining process includes removal of the overburden (i.e., primary soils and vegetation), excavation of the resource, and transportation to a processing facility. Mining accounts for slightly more than 50% of current production, and is expected to remain 40%-50% through 2030.[10]
- **In-Situ.** Oil sands deposits that are deeper than approximately 75 meters are recovered using in-situ methods. Most in-situ recovery methods currently in operation involve injecting steam into an oil sands reservoir to heat—and thus decrease the viscosity of—the bitumen, enabling it to flow out of the reservoir to collection wells. Steam is injected using cyclic steam stimulation (CSS), where the same well cycles both the steam and the bitumen, or by steam-assisted gravity drainage (SAGD), where a top well is used for steam injection and the bottom well is used for bitumen recovery. Because significant amounts of energy are currently required to create steam, in-situ methods are generally more GHG-intensive than conventional mining (excluding land use impacts). With over 80% of recoverable reserves

situated too deep for conventional mining techniques, it is assumed that the industry will eventually move toward an increased use of the in-situ extraction process in some form.

Study Design Factors. Design factors relate to how the GHG comparison is structured in each study and which parameters are included. These factors may include:

- overall purpose and goal of the study,
- time frame for the inputs and the results,
- life-cycle boundaries that are established for comparison,
- units and metrics used for comparison,
- GHG global-warming potential used for comparison,[11]
- treatment of co-products during refining (e.g., asphalt, petroleum coke, liquid gases, lubricants),
- treatment of secondary emission flows (e.g., capital infrastructure, land-use changes),[12]
- treatment of power co-generation at the facilities, and
- treatment of flaring, venting, and fugitive emissions.

Input Assumptions. Input assumptions can impact life-cycle results at each stage of the assessment. Studies often use simplified assumptions to model GHG emissions due to limited data availability and the complexity of and variability in the practices used to extract, process, refine, and transport crude oil, diluted crude, or refined product. Key input assumptions for Canadian oil sands crude may include:

- percentage contribution of each type of crude and each type of extraction process in the final transported product,
- type of upgrading or refining processes,
- amount of petroleum coke produced, stored, combusted, or sold,
- steam-to-oil ratio for in-situ extraction,
- bitumen-to-diluents ratio for dilbit, and
- energy efficiency of steam generation and other production processes.

Results of Selected Life-Cycle Emissions Assessments

Life-Cycle Assessments of Canadian Oil Sands

Greenhouse gases, primarily in the form of carbon dioxide and methane, are emitted during a variety of stages in oil sands production (see text box below for a summary).[13]

Summary of the Potential Sources of GHG Emissions in Oil Sands Development

- land use changes (emissions from the removal of vegetation and trees, soil, and peatland for mining or facilities),
- capital equipment (emissions from the construction of facilities, machinery, or other infrastructure),
- upstream fuels (emissions from the upstream production of fuel or electricity that is imported to the facility to be used as process heat or power for machinery),
- extraction (emissions from the bitumen extraction process, including equipment for mining and steam generation for artificial lifting),
- upgrading (emissions from the bitumen upgrading process and the combustion of co-products),
- crude product transportation (emissions from the transportation of crude products and co-products),
- refining (emissions from the crude oil refining process and the combustion of co-products),
- fugitives (emissions from the venting or flaring of methane, or fugitive leaks at any stage of production),
- refined product transportation (emissions from the transportation of final refined products and co-products),
- combustion (emissions from the end-use combustion of the refined fuel and co-products).

Table 1. Life-Cycle Assessments of Canadian Oil Sands As evaluated by DOS/IFC for inclusion in the Keystone XL Project Final EIS.

Study	Reference Years	Type	Boundaries	Design Factors
	Primary LCAs, the data from which are included in the Final EIS			
AERI/Jacobs 2009	2000s	LCA	WTW	All crudes
AERI/TIAX 2009	2007-2009	LCA	WTW	All crudes
NETL 2008	2005	LCA	WTW	All crudes
NETL 2009	2005	LCA	WTW	All crudes
	Other studies, the data from which are not included in the Final EIS			
Charpentier 2009	1999-2008	Meta-analysis	WTW	Dilbit not analyzed
GREET 2010	Current	Model	WTW	SCO and dilbit unspecified
ICCT 2010	2009	Partial LCA	WTT	Only average mix of imports to Europe analyzed
IEA 2010	2005-2009	Meta-analysis	WTW	Crude type not specified, results compared on a per barrel basis
IHS CERA 2010	2005-2030	Meta-analysis	WTW	All crudes, results compared on a per barrel basis
McCann 2001	2007	LCA	WTW	SCO only, results compared on a per liter basis
McCulloch/Pembina 2006	2002-2005	Partial LCA	WTR	SCO only, results compared on a per barrel basis
NRCan 2008	2008	LCA	WTW	Bitumen only, dilbit not analyzed
NRDC 2010	2006-2010	Meta-analysis	WTW	All crudes
Pembina 2005	2000, 2004	Partial LCA	WTR	Crude composition not specified
RAND 2008	2000s	LCA	WTR	SCO only

Sources: Jacobs Consultancy, Life Cycle Assessment Comparison of North American and Imported Crudes, Alberta Energy Research Institute and Jacobs Consultancy, 2009; TIAX LLC, Comparison of North American and Imported Crude Oil Lifecycle GHG Emissions, Alberta Energy Research Institute and TIAX LLC, 2009; NETL, Development of Baseline Data and Assessment of Life Cycle Greenhouse Gas Emissions of Petroleum-Based Fuels, National Energy Technology Laboratory, November 26, 2008; NETL, An Evaluation of the Extraction, Transport and Refining of Imported Crude Oils and the Impact of Life Cycle Greenhouse Gas

Emissions, National Energy Technology Laboratory, March 27, 2009; Charpentier, A.D., et al., "Understanding the Canadian Oil Sands Industry's Greenhouse Gas Emissions," Environmental Research Letters, Vol. 4, January 20, 2009; GREET, Greenhouse Gases, Regulated Emissions, and Energy Use in Transportation Model, Version 1.8d.1, Argonne National Laboratory, 2010; ICCT, Carbon Intensity of Crude Oil in Europe Crude, International Council on Clean Transportation, 2010; IEA, World Energy Outlook 2010, International Energy Agency, 2010; IHS CERA, Oil Sands, Greenhouse Gases, and U.S. Oil Supply: Getting the Numbers Right, IHS Cambridge Energy Research Associates, Inc., 2010; McCann and Associates, Typical Heavy Crude and Bitumen Derivative Greenhouse Gas Life Cycles in 2007, Prepared for Regional Infrastructure Working Group by T. J. McCann and Associates Ltd., November 16, 2001; McCulloch, M., et al., Carbon Neutral 2020: A Leadership Opportunity in Canada's Oil Sands, Oil Sands Issue Paper No. 2, Pembina Institute, October 2006; NRCan/$(S\&T)^2$, 2008 GHGenius Update, $(S\&T)^2$ Consultants report on model results prepared for Natural Resources Canada, August 15, 2008; NRDC, GHG Emission Factors for High Carbon Intensity Crude Oils, Ver. 2, Natural Resources Defense Council, September 2010; Pembina Institute, Oil Sands Fever: The Environmental Implications of Canada's Oil Sands Rush, November 2005; RAND Corporation. Unconventional Fossil-Based Fuels: Economic and Environmental Trade-Offs, The RAND Corporation, 2008; U.S. Department of State, Keystone XL Project, Final Environmental Impact Statement, Appendix V, "Life-Cycle Greenhouse Gas Emissions of Petroleum Products from WCSB Oil Sands Crudes Compared with Reference Crudes," July 13, 2011.

Notes: According to the DOS/IFC evaluation: "Type" is considered sufficient when the study is a unique, original assessment, and is not a meta-analysis that summarizes and averages the results from other sources; "Boundaries" is considered sufficient when the study evaluates the full WTW GHG emissions life cycle; "Design Factors" is considered sufficient when the study includes and evaluates all crude types likely to be transported by the Keystone XL pipeline. See DOS Final EIS, p. 40, for more on the DOS evaluation of each study.

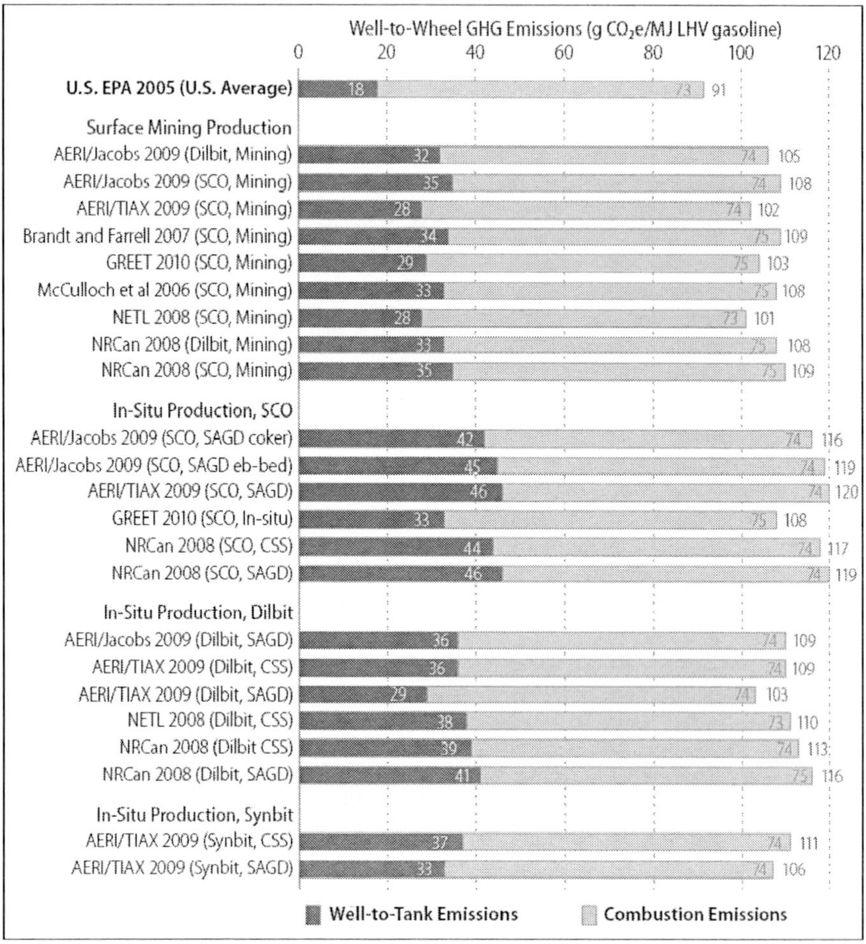

Source: CRS, from studies outlined in Table 1. Average U.S. petroleum baseline for 2005 provided by U.S. Environmental Protection Agency (U.S. EPA), *Renewable Fuel Standard Program (RFS2): Regulatory Impact Analysis*, February 2010, EPA-420-R-10-006, with data sourced from DOE/NETL, *Development of Baseline Data and Analysis of Life Cycle GHG Emissions of Petroleum Based Fuels*, November 2008.

Notes: See section "Life-Cycle Assessment Methodology" for key to crude oil types and production processes. U.S. EPA 2005 (U.S. average) assesses a weighted average of all imported transportation fuels to the United States in 2005, including Canadian oil sands. It does not include emissions from some of the most carbon-intensive imported crude oils (e.g., Venezuelan Heavy) due to modeling uncertainties.

Figure 2. Well-to-Wheel GHG Emissions Estimates for Canadian Oil Sands Crudes.

Table 2. Reported Findings of Well-to-Wheel GHG Emissions Estimates in the Life-Cycle Assessments of Canadian Oil Sands Crudes

Study	Production Method	Crude Type	WTW GHG Emissions	Increase over Baseline	Key Assumptions
		LCAs analyzed by IFC 2011			
	WTW GHG emissions expressed in gCO2e/MJ LHV gasoline				
U.S. EPA 2005	Baseline	Varied	91	—	Baseline assesses the emissions from a weighted average of imported transportation fuels to the United States in 2005, including Canadian oil sands. It does not, however, include emissions from some of the most carbon-intensive imported crude oils (e.g., Venezuelan Heavy) due to modeling uncertainties.
AERI/Jacobs 2009	Mining + Upgrading	SCO	108	19%	Units: gCO2e/MJ reformulated gasoline; petroleum coke stored at upgrader, allocated to other fuel products outside LCA at refinery; accounting for upgrading included in refinery emissions; emissions from upstream fuel production included; venting and flaring included; infrastructure and land-use changes not specified or not included.
AERI/Jacobs 2009	Mining	Dilbit	105	15%	Units: gCO2e/MJ reformulated gasoline; diluents processed with bitumen at refinery; emissions from upstream fuel production included; venting and flaring included; infrastructure and land-use changes not specified or not included.
AERI/Jacobs 2009	In-Situ, SAGD + Upgrading (Hydrocracking)	SCO	119	31%	Units: gCO2e/MJ reformulated gasoline; steam-to-oil ratio (SOR) of 3; petroleum coke stored at upgrader, allocated to other fuel products outside LCA at refinery; cogeneration credits applied; accounting for upgrading included in refinery emissions; emissions from upstream fuel production included; venting and flaring included; infrastructure and land-use changes not specified or not included.

Table 2. (Continued)

Study	Production Method	Crude Type	WTW GHG Emissions	Increase over Baseline	Key Assumptions
AERI/Jacobs 2009	In-Situ, SAGD + Upgrading (Coker)	SCO	116	27%	Units: gCO2e/MJ reformulated gasoline; SOR 3; petroleum coke stored at upgrader, allocated to other fuel products outside LCA at refinery; cogeneration credits applied; accounting for upgrading included in refinery emissions; emissions from upstream fuel production included; venting and flaring included; infrastructure and land-use changes not specified or not included.
AERI/Jacobs 2009	In-Situ, SAGD	Dilbit	105-113	15%-24%	Units: gCO2e/MJ reformulated gasoline; SOR 3; cogeneration credits applied; diluents processed with bitumen at refinery; emissions from upstream fuel production included; venting and flaring included; infrastructure and land-use changes not specified or not included.
AERI/TIAX 2009	Mining + Upgrading	SCO	102	12%	Units: gCO2e/MJ reformulated gasoline; petroleum coke not combusted at upgrader, allocated to other fuel products outside LCA at refinery; accounting for upgrading included in refinery emissions; emissions from upstream fuel production included; venting, flaring, and fugitives included; infrastructure and land-use changes not specified or not included.
AERI/TIAX 2009	In-Situ, SAGD + Upgrading	SCO	112-128	23%-41%	Units: gCO2e/MJ reformulated gasoline; SOR 2.5; petroleum coke not combusted at upgrader, allocated to other fuel products outside LCA at refinery; cogeneration credits applied using project specific data; accounting for upgrading included in refinery emissions; emissions from upstream fuel production included; venting, flaring, and fugitives included; infrastructure and land-use changes not specified or not included.

Study	Production Method	Crude Type	WTW GHG Emissions	Increase over Baseline	Key Assumptions
AERI/TIAX 2009	In-Situ, SAGD	Synbit	105-108	15%-19%	Units: gCO2e/MJ reformulated gasoline; SOR 2.5; cogeneration credits applied using project specific data; emissions from upstream fuel production included; venting, flaring, and fugitives included; infrastructure and land-use changes not specified or not included.
AERI/TIAX 2009	In-Situ, SAGD	Dilbit	101-105	11%-15%	Units: gCO2e/MJ reformulated gasoline; SOR 2.5; cogeneration credits applied using project specific data; diluents processed with bitumen at refinery; emissions from upstream fuel production included; venting, flaring, and fugitives included; infrastructure and land-use changes not specified or not included.
AERI/TIAX 2009	In-Situ, CSS	Synbit	109-112	20%-23%	Units: gCO2e/MJ reformulated gasoline; SOR 3.4-4.8; cogeneration credits applied using project specific data; emissions from upstream fuel production included; venting, flaring, and fugitives included; infrastructure and land-use changes not specified or not included.
AERI/TIAX 2009	In-Situ, CSS	Dilbit	107-112	18%-23%	Units: gCO2e/MJ reformulated gasoline; SOR 3.4-4.8; cogeneration credits applied using project specific data; diluents processed with bitumen at refinery; emissions from upstream fuel production included; venting, flaring, and fugitives included; infrastructure and land-use changes not specified or not included.
NETL 2008	Mining + Upgrading	SCO	101	11%	Units: gCO2e/MMBtu gasoline, diesel, and jet fuel; petroleum coke use unspecified at upgrader, allocated outside LCA at refinery; accounting for upgrading not included in refinery emissions; emissions from upstream fuel production included; venting, flaring, and fugitives included; infrastructure and land-use changes not specified or not included.

Table 2. (Continued)

Study	Production Method	Crude Type	WTW GHG Emissions	Increase over Baseline	Key Assumptions
NETL 2008	In-Situ, CSS	Dilbit	110	21%	Units: gCO2e/MMBtu gasoline, diesel, and jet fuel; SOR not stated; cogeneration unspecified; diluents unspecified; emissions from upstream fuel production included; venting, flaring, and fugitives included; infrastructure and land-use changes not specified or not included.
		Additional LCAs analyzed by NRDC 2010 WTW GHG emissions expressed in gCO_2e/MJ LHV gasoline			
U.S. EPA 2005	Baseline	Varied	93	—	Baseline assesses a weighted average of all imported transportation fuels to the United States in 2005. Includes emissions from higher carbon-intensity crude oils imported or produced domestically.
GREET 2010	Mining + Upgrading	SCO	103	11%	Units: gCO2e/mile; petroleum coke use unspecified; accounting for upgrading not included in refinery emissions; emissions from upstream fuel production not specified; venting, flaring, and fugitives included; infrastructure and land-use changes not specified or not included.
GREET 2010	In-Situ, SAGD + Upgrading	SCO	108	16%	Units: gCO2e/mile; SOR not stated; petroleum coke use unspecified; cogeneration unspecified; accounting for upgrading not included in refinery emissions; emissions from upstream fuel production not specified; venting, flaring, and fugitives included; infrastructure and land-use changes not specified or not included.
McCulloch 2006	Mining + Upgrading	SCO	105-111	13%-19%	Units: kgCO2e/bbl SCO; petroleum coke gasification at upgrader included in high estimate, unspecified at the refinery; accounting for upgrading not specified in refinery emissions; emissions from upstream fuel production not specified; venting, flaring, and fugitives partially included; infrastructure and land-use changes not specified or not included.

Study	Production Method	Crude Type	WTW GHG Emissions	Increase over Baseline	Key Assumptions
NRCan 2008	Mining + Upgrading	SCO	109	17%	Units: gCO2e/MJ reformulated gasoline; petroleum coke used at the upgrader contributes 15% of the energy requirement for processing SCO and the remainder offsets emissions from coal combustion at electric generating units, not specified at refinery; accounting for upgrading not included in refinery emissions; emissions from upstream fuel production included; venting, flaring, and fugitives included; infrastructure and land-use changes not specified or not included.
NRCan 2008	Mining	Dilbit	108	16%	Units: gCO2e/MJ reformulated gasoline; diluents unspecified; emissions from upstream fuel production included; venting, flaring, and fugitives included; infrastructure and land-use changes not specified or not included.
NRCan 2008	In-Situ, SAGD + Upgrading	SCO	119	28%	Units: gCO2e/MJ reformulated gasoline; SOR 3.2; petroleum coke used at the upgrader contributes 15% of the energy requirement for processing SCO and the remainder offsets emissions from coal combustion at electric generating units, not specified at refinery; cogeneration not included; accounting for upgrading not included in refinery emissions; emissions from upstream fuel production included; venting, flaring, and fugitives included; infrastructure and land-use changes not specified or not included.
NRCan 2008	In-Situ, SAGD	Dilbit	116	25%	Units: gCO2e/MJ reformulated gasoline; SOR 3.2; cogeneration not included; diluents unspecified; emissions from upstream fuel production included; venting, flaring, and fugitives included; infrastructure and land-use changes not specified or not included.
NRCan 2008 NRCan 2008	In-Situ, CSS + Upgrading In-Situ, CSS	SCO Dilbit	117 113	26% 22%	Units: gCO2e/MJ reformulated gasoline; SOR not stated; petroleum coke used at the upgrader contributes 15% of the energy requirement for processing SCO and the remainder offsets emissions from coal combustion at electric generating units, not specified at refinery; cogeneration not included; accounting for upgrading not included in refinery emissions;

Table 2. (Continued)

Study	Production Method	Crude Type	WTW GHG Emissions	Increase over Baseline	Key Assumptions
					emissions from upstream fuel production included; venting, flaring, and fugitives included; infrastructure and land-use changes not specified or not included. Units: gCO2e/MJ reformulated gasoline; SOR not stated; cogeneration not included; diluents unspecified; emissions from upstream fuel production included; venting, flaring, and fugitives included; infrastructure and land-use changes not specified or not included.
	Additional LCAs analyzed by IHS CERA 2010 WTW GHG emissions expressed in KgCO2e/barrel of refined product (see notes below)				
IHS CERA, 2010	Average US Barrel Consumed	Varied	487	—	As modeled by IHS CERA from data sourced from NETL 2008.
IHS CERA, 2010	Mining	Dilbit	488	<1%	Units: $kgCO_2e$ per barrel of refined products; diluents processed with bitumen at refinery; emissions from upstream fuel production not included; venting, flaring, and fugitives not specified; infrastructure and land-use changes not specified or not included.
IHS CERA, 2010 IHS CERA, 2010	Mining + Upgrading (Coker) In-Situ, SAGD	SCO Dilbit	518 512	6% 5%	Units: $kgCO_2e$ per barrel of refined products; petroleum coke use unspecified at the upgrader, allocated outside LCA at refinery; accounting for upgrading not specified in refinery emissions; emissions from upstream fuel production not included; venting, flaring, and fugitives not specified; infrastructure and land-use changes not specified or not included. Units: $kgCO_2e$ per barrel of refined products; SOR 3; cogeneration credits applied; diluents processed with bitumen at refinery; emissions from upstream fuel production not included; venting, flaring, and fugitives not specified; infrastructure and land-use changes not specified or not included.

Study	Production Method	Crude Type	WTW GHG Emissions	Increase over Baseline	Key Assumptions
IHS CERA, 2010	In-Situ, SAGD + Upgrading	SCO	555	14%	Units: $kgCO_2e$ per barrel of refined products; SOR 3; petroleum coke use unspecified at the upgrader, allocated outside LCA at refinery; cogeneration credits applied; accounting for upgrading not specified in refinery emissions; emissions from upstream fuel production not included; venting, flaring, and fugitives not specified; infrastructure and land-use changes not specified or not included.

Sources: CRS, from studies outlined in Table 1. Average U.S. petroleum baseline for 2005 provided by U.S. EPA, *Renewable Fuel Standard Program (RFS2): Regulatory Impact Analysis*, February 2010, EPA-420-R-10-006, with data sourced from DOE/NETL, *Development of Baseline Data and Analysis of Life Cycle GHG Emissions of Petroleum Based Fuels*, November 2008.

Notes: See section "Life-Cycle Assessment Methodology" for key to crude oil types and production processes. IFC 2011 and the LCAs it reviewed, as well as NRDC 2010, expressed functional units in GHG emissions per megajoule (MJ) of gasoline, per MJ of diesel, and per MJ of jet fuel (the gasoline values are shown in this report). IHS CERA 2010, in contrast, expressed GHG emissions in units of kilograms of carbon dioxide equivalent per barrel of refined product produced, ($kgCO_2e$ per barrel of refined products). Refined products are defined by IHS CERA as "the yield of gasoline, diesel, distillate, and gas liquids from each crude." As a meta-analysis, IHS CERA 2010 used the results of the existing and publicly available life-cycle assessments, including many of those listed in Table 1; however, a demonstration of the unit conversions was not provided. Without detail of the underlying allocation methods used to aggregate the gasoline, diesel, jet fuel, and other co-products, neither CRS nor the DOS/IFC report was able to convert and directly compare IHS CERA's functional units to the other studies.

A number of published and publicly available studies have attempted to assess the life-cycle GHG emissions data for Canadian oil sands crudes. This report examines the life-cycle assessments analyzed by the U.S. Department of State (DOS)—in conjunction with the consultancy firm IFC International LLC (IFC)—in the Keystone XL Project's Final Environmental Impact Statement (Final EIS).

The studies were selected by IFC using several criteria: (1) they evaluated Canadian oil sands crudes in comparison to other reference crude oils, (2) they focused on GHG emissions impacts throughout the entire crude oil life-cycle, (3) they were published within the past 10 years, and (4) they represented the perspectives of a range of stakeholders.

Table 1 provides a list of the studies referenced by the IFC analysis. While the type, boundaries, and design features vary across all studies, DOS and IFC determined the data and results from AERI/Jacobs 2009, AERI/TIAX 2009, NETL 2008, and NETL 2009 to be sufficiently robust for inclusion in the Final EIS. Reasons against the inclusion of the remaining studies are presented briefly in the table, and outlined in more detail in the Final EIS.

Findings

Using data from various studies, the DOS/IFC analysis in the Final EIS finds the following:

- Well-to-Wheel GHG emissions of 91 gCO_2e/megajoule (MJ) lower heating value (LHV)[14] gasoline for the average of imported transportation fuels to the United States in the reference year of 2005.[15]
- Well-to-Wheel GHG emissions of, on average, 104-109 gCO_2e/MJ LHV gasoline for the weighted average[16] of Canadian oil sands crudes likely to be transported in the proposed Keystone XL pipeline project in the near term.
- An increase in Well-to-Wheel GHG emissions of 14%-20% for Canadian oil sands crude over the 2005 average for all imported transportation fuels to the United States.
- An increase in Well-to-Tank (i.e., "production") GHG emissions of 72%-111% for Canadian oil sands crude over the 2005 average production emissions for imported transportation fuels to the United States (18 gCO_2e/MJ).

Individual estimates of WTW GHG emissions from Canadian oil sands crude types and processes from the secondary studies listed in Table 1 range from increases of 1%-41% over the baseline.

Figure 2 presents a summary of the WTW GHG emissions estimates for various Canadian oil sands crude types and production processes as reported by each study. Table 2 summarizes and compares each study's emissions estimates, data, and relevant input assumptions used to identify the key drivers in the life-cycle GHG emissions. Variability among the estimates is the result of each study's differing design and input assumptions. A discussion of these assumptions—and their estimated effects on GHG emissions impacts—follows in the next section.

Design Factors and Input Assumptions for Canadian Oil Sands Assessments

Most published and publicly available studies on the life-cycle GHG emissions data for Canadian oil sands identify two main factors contributing to the increase in emissions relative to other reference crudes:

1) oil sands are heavier and more viscous than lighter crude oil types on average, and thus require more energy- and resource-intensive activities to extract;
2) oil sands are compositionally deficient in hydrogen, and have a higher carbon, sulfur, and heavy metal content than lighter crude oil types on average, and thus require more processing to yield consumable fuels.

While most studies agree that Canadian oil sands crudes are on average "somewhat" more GHGintensive than the crudes they may displace in the U.S. refineries, the range of the reported increase varies among assessments. Key design and input assumptions can significantly influence results. These factors include:

- **Metrics.** Comparing results from various studies is complicated by each study's choice of functional units. While GHG emissions have been normalized by most studies and reported as CO2-equivalents, the units they are expressed "over" vary greatly. Some evaluate GHG emissions on the basis of a particular final fuel product (e.g., gasoline, diesel, or jet fuel). Others evaluate emissions by an averaged barrel of refined product. Some studies report emissions per unit of volume (e.g., millions of barrels (mbl)), and others by unit of energy produced (e.g., British Thermal Units (Btus) or megajoules (MJ)). For example, NETL 2008, Jacobs 2009, and TIEX 2009 use functional units for energy produced across the final products—MMBtus or MJs for gasoline, diesel, and/or jet fuel. IHS CERA 2010 expresses GHG emissions "per barrel of refined product produced"; while others, like Charpentier 2009 (not included in the reported findings), by "kilometers driven." The choice affects how the results are presented and makes it challenging to compare across studies if the data or conversion values are not fully published or transparent.

- **Extraction Process.** GHG emissions vary by the type of extraction process used to recover bitumen. Due to the high energy demands of steam production, in-situ methods are generally assumed to be more GHG-intensive than mining operations. However, not all studies assess the difference to be the same. IHS CERA 2010 estimates the increase of WTW GHG emissions from in-situ extraction to be, on average, 7% greater than mining. NRDC 2010 estimates 9%. Specific estimates in Jacobs 2009 show a 4% increase (for SAGD dilbit over mining dilbit) and in NRCan 2008 an increase of 9% (for SAGD SCO over mining SCO).
- **In-Situ Steam-to-Oil (SOR) Ratio.** The amount of steam injected into a reservoir during in-situ processes to extract a unit volume of bitumen varies across reservoirs and across extraction facilities. The resulting energy consumption and GHG emissions estimates vary accordingly.
- Thus, the figure used in LCAs to express this ratio may significantly impact GHG estimates. NRCan 2008 reports SOR values from 2.5 to 5.0 across SAGD operations in Canadian oil sands. NRDC 2010 reports a range from 1.94 to 7.26. IHS CERA cites an industry average of 3. Charpentier 2009 demonstrates that GHG emissions at the production phase are very sensitive to SOR, estimating that every 0.5 increase in the ratio corresponds to an increase of 10 kgCO2e GHG emissions per barrel of bitumen produced.
- **Upgrading Process.** Bitumen needs pre-processing in order to lower its viscosity and remove impurities before it is fit for conventional refineries. This preprocessing is called "upgrading," the key components of which include (1) removal of water, sand, physical waste, and lighter products; (2) catalytic purification (i.e., the process of removing excess sulfur, oxygen, nitrogen, and metals), and (3)

hydrogenation through either carbon rejection or catalytic hydrocracking (i.e., the process of removing or breaking down the heaviest fraction of the oil residuum by either vacuum distillation and precipitation or by adding hydrogen in a "hydrocracking process that breaks long-chain hydrocarbons into shorter, more useful ones). The residuum can be further refined in a "coking" process to produce gasoline, distillate, and petroleum coke.

- The resulting product is synthetic crude oil (SCO) and numerous co-products, including water, sand, waste, sulfur, oxygen, nitrogen, distillate, and petroleum coke, among others. Some of the co-products from the upgrading process contain carbon and other potential GHG emission sources. Thus, a consistent and comprehensive accounting of the GHG emission from all co-products would be necessary for a full life-cycle assessment of oil sands crude—or any hydrocarbon—production.
- **Treatment of Petroleum Coke.** Petroleum coke (an excess source of carbon) is a co-product of bitumen production at both the upgrader and the refinery.
- Roughly 5%-10% of a barrel of crude ends up as coke; and the heavier the crude, the greater the percentage of coke. Dilbit refining produces about 50% more coke than the average conventional crude. The treatment of coke is a primary driver behind the results of any WTW GHG oil sands assessment. If coke is combusted (i.e., for process heat, electricity, or hydrogen production *at the upgrader* in lieu of natural gas combustion), WTW GHG emissions may increase anywhere from 14% (TIAX 2009) to 50% (McCulloch 2006) over lighter crudes. If it is stored, sold, and/or combusted elsewhere, its potential emissions may not be factored into the LCA. The main concern for modeling is ensuring that coke produced at the upgrader (for SCO) is treated consistently with coke produced at the refinery (for dilbit or other imported crudes). Based on the studies analyzed in this report, petroleum coke at the upgrader is either (1) combusted (increase in WTW GHG emissions), (2) stored (unspecified change in WTW GHG emissions), or (3) sold as a fuel for combustion (increase in WTW GHG emission if allocated to coke production (as in TIAX 2009)). In contrast, petroleum coke at the refinery is either (1) used as an offset credit to back out coal combustion for electricity generation (decrease in WTW GHG emissions), or (2) allocated outside the life-cycle assessment (effect not included in WTW GHG emissions).

These inconsistent methodologies make comparisons problematic. Coke produced at U.S. refineries has a low domestic demand, and is therefore often shipped to overseas markets for use as a replacement fuel for coal combustion or steel production (most studies include neither the overseas transportation nor the combustion emissions of coke in WTW GHG emissions assessments).

- **Cogeneration.** Cogeneration facilities use both steam and electricity generated from the steam to achieve higher energy efficiencies. In-situ extraction facilities often have steam requirements much greater than electricity requirements, thus leaving excess capacity for electricity generation that can be exported back into the grid for use elsewhere. Offset credits given to exported electricity in LCAs can have a substantial impact on WTW GHG emissions. Cogeneration assumptions vary across the studies of Canadian oil sands crudes in two ways: (1) whether cogeneration credits are included, and (2) if so, what source of electricity is offset (e.g., coal-fired generation, oil, or natural gas). Some estimates show that applying credits from oil sands facilities to offset coal-fired electricity generation could reduce WTW GHG emission to within the range of conventional crudes. Many studies currently do not consider offset credits because the practice is not in widespread use among producers.
- **Upgrading and Refinery Emissions.** Because SCO delivered to a refinery has already been processed at the upgrader, the energy consumption at the refinery— and therefore the GHG emissions at the refinery—will be lower than the refinery emissions of dilbit or other crudes. Accounting for the reduced emissions from SCO has a modest effect on WTW GHG emissions, as refinery emissions are commonly around 5%-15% of the total. Many studies do not mention this accounting, and it is unclear if the reductions for SCO at the refinery are incorporated into many of the LCAs.
- **Diluents.** Because the viscosity of raw bitumen is too high to be transported via pipeline, diluting bitumen with lighter hydrocarbons to assist in its transport has become a common industry practice. Accounting for the effects of diluting bitumen is an important component in emission estimates, because producing and refining the diluents into finished products may result in a lower WTW GHG emissions estimate *per barrel* of dilbit in comparison to a barrel of raw bitumen. LCAs that report emissions for dilbit on a per barrel of refined product basis (e.g., IHS CERA 2010) are thus reporting the

emissions from a combination of both oil sands bitumen and the supplemental hydrocarbons. Additionally, diluting raw bitumen with light hydrocarbons creates a crude product that is more difficult and energy-intensive to refine than other crude oils, thus producing less premium refined product per barrel after the refinery stage.[17] The extent to which this difference in yield is accounted for across the various studies is unclear. The IHS CERA 2010 estimates for crude production of SAGD dilbit do not show an adjustment for the difference. TIAX 2009 and Jacobs 2009 both show slightly higher refinery emissions for dilbit compared to other crudes, but the reasons for the increase are not specified.

- **Upstream Production Fuels.** Some studies include the GHG emissions associated with the upstream production of purchased electricity that is imported to provide process heat and to power machinery throughout crude production. The upstream GHG emissions for natural gas fuel and electricity generation used in the production of oil sands can be significant. Jacobs 2009 demonstrates that the GHG emissions associated with the upstream fuel cycle account for roughly 4%-5% of the total WTW GHG emissions for average Canadian oil sands. IHS CERA 2010 indicates that although its study excludes upstream fuel and electricity GHG emissions, the inclusion of them would add 3% to WTW GHG emissions per barrel of refined product.
- **Flared, Vented, and Fugitive Emissions.** Emissions associated with flaring and venting can be a significant source of GHG emissions. The TIAX 2009 study indicates that including venting and flaring emissions associated with oil sands production (particularly for mining extraction techniques) may contribute up to 4% of total WTW GHG emissions. Further, methane emissions from fugitive leaks throughout the oil sands production process can potentially contribute up to 1% of GHG emissions.[18] Methane emissions from oil sands mining and tailings ponds may have an even larger impact, contributing from 0%-9% of total GHG emissions.[19] TIAX 2009, McCulloch 2006, and NRCan 2008 state that they include emissions from these sources. IHS CERA 2010 excludes emissions from methane released from tailings ponds but recognizes there is considerable uncertainty and variance in quantifying these emissions. Other studies do not specify.

- **Infrastructure/Construction Emissions.** None of the existing studies include the GHG impacts associated with capital equipment and the construction of facilities, machinery, and infrastructure needed to produce oil sands. According to Bergerson and Keith 2006,[20] the relative percentage increase to WTW GHG emissions from incorporating capital equipment can be between 9% and 11%. Charpentier 2009 discusses the need to more fully investigate and include these potentially significant supply chain infrastructure GHG emissions in future oil sands life-cycle studies.
- **Local and Indirect Land-Use Change Emissions.** Emissions associated with changes in biological carbon stocks from the removal of vegetation, trees, and soil during oil sands mining operations may be significant, albeit temporary. Yeh 2010 estimates that surface mining of oil sands results in a 0.9%-2.5% increase in the WTW emissions versus the baseline (2005 average U.S. gasoline). The range was highly dependent on the type of lands displaced and reclamation practices used, with the removal of peatland having the largest impact. None of the life-cycle studies reviewed, however, include land-use change GHG emissions in the WTW life-cycle assessment. Several recent studies have begun to assess the effects.[21]

Life-Cycle Assessments of Canadian Oil Sands versus Other Reference Crudes

To compare the life-cycle GHG emissions from Canadian oil sands crudes against those of other crude oils imported into the United States, many of the published studies conduct reference assessments of other global resources.

Findings

Figure 3 presents the results of one of the more comprehensive studies (NETL 2009), which compares Well-to-Wheel GHG emissions of reformulated gasoline across various crude oil feedstocks (a review of the NETL 2009 input assumptions is included in the figure's "Notes" section). The NETL findings show the following:

- Well-to-Wheel GHG emissions from gasoline produced from a weighted average of Canadian oil sands crudes imported to the United States are approximately 17% higher than that from gasoline derived from the average mix of crudes imported to the United States in 2005. This corresponds to an increase in Well-toTank (i.e., "production") GHG emissions of 80% over the 2005 average production emissions for imported transportation fuels to the United States (18 gCO_2e/MJ).

- Well-to-Wheel GHG emissions from gasoline produced from a weighted average of Canadian oil sands crudes are 19%, 12%, and 18% higher than the life-cycle emissions from Middle Eastern Sour, Mexican Maya, and Venezuelan Conventional crudes, respectively.[22] This corresponds to an increase in Well-toTank (i.e., "production") GHG emissions of 102%, 53%, and 92% higher than the production emissions from Middle Eastern Sour, Mexican Maya, and Venezuelan Conventional crudes, respectively.

Individual estimates of WTW GHG emissions from Canadian oil sands crudes from the primary studies listed in Table 1 range from 9%-19% more GHG-intensive than Middle Eastern Sour, 5%-13% more GHG-intensive than Mexican Maya, and 2%-18% more GHG-intensive than various Venezuelan crudes (including Venezuelan Conventional and Bachaquero).

Design Factors and Input Assumptions for Reference Crudes Assessments

Similar to the LCAs conducted on Canadian oil sands crudes, assessments on other global resources present many variables and uncertainties in the available data. Likewise, these assessments are bounded by specific design factors and input assumptions that can affect the quality of the results. These factors include:

- **Choice of Reference Crudes Studied.** Crude oil resources around the world vary significantly in regard to resource quality and production methods. Thus, the comparison of the GHG emissions intensity for Canadian oil sands crudes against a reference crude could vary significantly, depending upon the choice of reference.
- **Choice of Reference Crudes Presumed Displaced in U.S. Refineries.** If an increased volume of Canadian crude is imported into the United States, and if U.S. refineries continue using the same input mix of heavy crudes as they currently use, it is assumed by some of

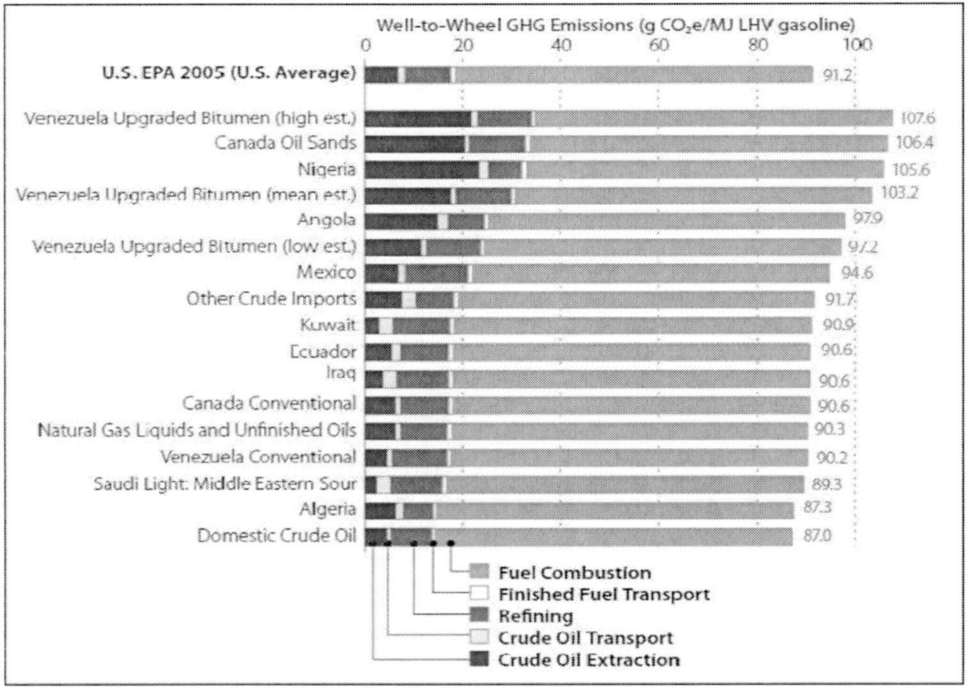

Source: CRS, from NETL, An Evaluation of the Extraction, Transport and Refining of Imported Crude Oils and the Impact of Life Cycle Greenhouse Gas Emissions, National Energy Technology Laboratory, March 27, 2009.

Notes: NETL's value for the final fuel consumption stage is 72.6 $kgCO_2e/MJ$ for all crude resources. U.S. EPA 2005 (U.S. average) assesses a weighted average of all imported transportation fuels to the United States in 2005, including Canadian oil sands. It does not include emissions from some of the most carbon-intensive imported crude oils (e.g., Venezuelan Heavy) due to modeling uncertainties. NETL values converted from kgCO_2e/MMBtu using conversion factors of 1,055 MJ/MMBtu and 1,000 g/kg. NETL input assumptions are as follows: (1) assumes a weighted average of Canadian oil sands extraction at 43% raw bitumen (not accounting for blending with diluents to form dilbit) from CSS in-situ production and 57% SCO from mining production in the years 2005 and 2006; (2) allocates refinery emissions from co-products other than the gasoline, diesel, and jet fuel to the co-products themselves, including petroleum coke, and thus outside the boundaries of the LCA (unless combusted at refinery); (3) uses linear relationships to relate GHG emissions from refining operations based on API gravity and sulfur content, thus failing to fully account for the various produced residuum ranges of bitumen blends and SCO; (4) does not fully evaluate the impact of pre-refining SCO at the upgrader prior to the refinery; (5) does not account for the transportation emissions of co-products; and (6) bounds the GHG emissions estimates for Venezuela's ultra-heavy oil/bitumen using uncertainty analysis due to the limited availability of public data. Further, as noted in Table 2, NETL 2009 study assumptions do not state SOR, do not include upstream fuel production, do not include infrastructure or land-use changes, and do not specify cogeneration, but do include emissions from venting, flaring, and fugitives.

Figure 3. Well-to-Wheel GHG Emissions Estimates for Global Crude Resources.

the studies that resources like Venezuelan Bachaquero or Mexican Maya would likely be the first displaced. However, to the extent that a crude like Saudi Light (i.e., Middle Eastern Sour) is the world's balancing crude, some analysts believe that it may ultimately be the resource that is backed out of the global market by increased Canadian oil sands production.[23]

- **Assumptions Regarding Artificial Lift.** Comparisons of reference crudes are further complicated by the choice of production phase, and thus the determination of the crude's recovery method. While primary resource recovery relies only upon the internal pressure of the well to lift the resource, the use of injected water or gas in subsequent phases of recovery, or the use of steam or CO_2 in the final phases, can significantly increase the emissions intensity of crude recovery.
- **Sensitivity to Water-Oil and Gas-Oil Ratios.** These ratios describe the fraction of the flow from a well that is oil, water, and gas. Due to the complex nature of production systems and resource reservoirs, studies often use ratios to develop simplified relationships between energy use and GHG emissions. Venting or flaring of associated gas, and fugitive emissions from produced water, can also have moderate to significant impacts on GHG emissions intensities.
- **Treatment of Petroleum Coke at the Refinery.** Much like the end-use determination—and consequent GHG emissions—for petroleum coke at Canadian upgrading facilities, the treatment of coke and its associated emissions at the refinery is a primary driver behind the results of any WTW GHG assessment of reference crudes.
- **Transportation Emissions.** LCAs must account for the contribution of transportation in WTW GHG emissions estimates, including the distance and mode of transportation from oil field to export terminal, and from producer to refiner, as well as the final transportation emissions of all co-products.

Life-Cycle Assessments of Canadian Oil Sands versus Other Fuel Resources

Figure 4 offers a comparison of the life-cycle GHG emissions estimates of petroleum products from Canadian oil sands crudes with estimates from other unconventional petroleum products, natural gas, and coal. These data are drawn from several different studies employing many different design features and input assumptions, not the least of which are different methods of combusting the final fuel products.

Further, it should be noted that different and non-substitutable end uses for the fuel products (e.g., the different end uses for coal and petroleum combustion) make a full comparison of their emissions impacts problematic. The figure presents an average value for each fuel; the original source materials provide a full description of each study's design characteristics as well as a presentation of each estimate's uncertainty analysis.

U.S. Carbon Footprint for the Keystone XL Pipeline

In response to comments on the draft Environmental Impact Statement (EIS) for the Keystone XL pipeline project and as a "matter of policy," the DOS/IFC study provides estimates for the incremental GHG emissions resulting in the production of Canadian oil sands crudes likely to be transported by the Keystone XL pipeline project (i.e., the U.S. GHG emissions footprint).[24]

Incremental GHG emissions are determined by the following:

1) the presumed throughput of the pipeline,
2) the mix of oil sands crude types imported through the pipeline, and
3) the GHG emission intensity of the crudes in the pipeline compared to the crudes they displace.

For illustrative purposes, the Final EIS uses the emissions assessment data from Jacobs 2009, TIAX 2009, and NETL 2009 to develop weighted averages of Canadian crudes and other reference crudes for the purposes of the carbon footprint analysis. DOS/IFC assumes the near-term initial throughput of the proposed Keystone XL pipeline to be 700,000 barrels of crude per day, with a potential capacity of 830,000 barrels per day. Using the Jacobs 2009 and TIAX 2009 assessments, DOS/IFC estimates that the throughput for the pipeline would be 50% SCO and 50% dilbit, with all dilbit produced using in-situ methods and 12% of SCO produced using in-situ methods, yielding a final mix of 50% in-situ produced dilbit, 44% mining produced SCO, and 6% in-situ produced SCO. Using the NETL 2009 assessment, DOS/IFC calculates a mix of 43% crude bitumen and 57% SCO. Incremental GHG emissions for

Canadian crudes are computed against four different reference crudes: Middle Eastern Sour (with the assumption that as the world's balancing crude, it may ultimately be the crude that is backed out of the world market by increased production of Canadian crudes), Mexican Maya and Venezuelan Bachaquero (with the assumption that as the heavy crudes currently in the input mix at U.S. refineries, they are likely to be the first displaced by an increased production of Canadian crudes), and a reference crude based on the average mix imported and refined in the United States in 2005.

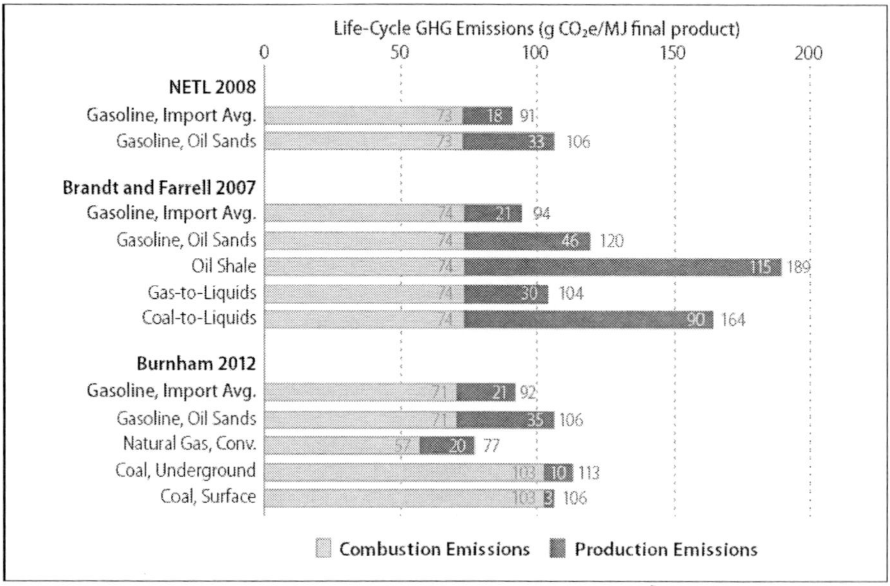

Source: CRS, from NETL, *Development of Baseline Data and Assessment of Life Cycle Greenhouse Gas Emissions of Petroleum-Based Fuels*, National Energy Technology Laboratory, November 26, 2008; Brandt, A.R. and A.E. Farrell, "Scraping the Bottom of the Barrel: Greenhouse Gas Emission Consequences of a Transition to Low-quality and Synthetic Petroleum Resources," *Climatic Change*, Vol. 84, 2007, pp. 241-263; and Burnham, A., et al., "Life-Cycle Greenhouse Gas Emissions of Shale Gas, Natural Gas, Coal, and Petroleum," *Environmental Science and Technology*, Vol. 46, 2012, pp. 619–627.

Notes: NETL values converted from kgCO2e/MMBtu using conversion factors of 1,055 MJ/MMBtu and 1,000 g/kg; Brandt values converted from gCe/MJ using conversion factor of 3.667 Ce/CO2e.

Figure 4. Life-Cycle GHG Emissions Estimates for Selected Fuel Resources/.

The analysis found that the potential range of incremental GHG emissions contributed by the pipeline would be 3-17 MMTCO2e annually at the near-term initial throughput and 4-21 MMTCO2e annually at the potential throughput. As the United States reported a total domestic GHG inventory of 6,865.5 MMTCO2e in 2010,[25] the incremental pipeline emissions would represent an increase of 0.06%-0.3% in total annual GHG emissions for the United States. This overall range is equivalent to annual GHG emissions from the combustion of fuels in approximately 588,000 to 4,061,000 passenger vehicles, or the CO_2 emissions from combusting fuels used to provide the energy consumed by approximately 255,000 to 1,796,000 homes for one year.[26]

FURTHER CONSIDERATIONS

Life-cycle assessment has emerged as an influential methodology for collecting, analyzing, and comparing the GHG emissions and climate change implications of various hydrocarbon resources. However, because of the complex life cycle of hydrocarbon fuels and the large number of analytical design features that are needed to model their emissions, LCAs retain many variables and uncertainties. These uncertainties often make comparing results across resources or production methods problematic.

Hence, the usefulness of LCA as an analytical tool for policymakers may lie less in its capacity to generate comparative rankings, or “scores,” between one source and another, and more in its ability to highlight “areas of concern,” or “hot spots,” in the production of a given hydrocarbon fuel. In this way, LCA can serve to direct policymakers’ attention to those areas in resource development that present the greatest challenges to GHG emissions control, and hence, the biggest potential benefits if adequately managed.

Table 3 summarizes the GHG emissions impacts of the various stages of Canadian oil sands production and presents examples of mitigation strategies that have been offered by industry, academia, and other stakeholders.

Table 3. Potential GHG Mitigation Activities in Canadian Oil Sands Production

Magnitude of Source's GHG Impact	Source of GHG	Mitigation Activity
Significant	Upstream Fuels for Production	Energy-efficiency measures. Use of natural gas or bio-based fuels such as biodiesel or bioethanol in mining and trucking fleets and equipment.
	Extraction	In-situ extraction improvements such as improved well configuration and placement, low-pressure SAGD, flue gas reservoir re-pressurization, new artificial lift pumping technologies, use of electric submersible pumps, and overall improvements in energy efficiency that can reduce the steam-to-oil ratios (SOR) of in-situ production processes. Steam solvent processes, which use solvents to reduce the steam required for bitumen extraction. These include solvent-assisted processes (SAP), expanding solvent steam-assisted gravity drainage (ES-SAGD), and liquid addition to steam for enhanced recovery (LASER). Electrothermal extraction, where electrodes are used to heat the bitumen in the reservoir.
		Use of lower-temperature water to separate bitumen from sand during extraction to reduce the energy required. In-situ combustion, where the heavy portion of petroleum is combusted underground.
	Upgrading and Refining	Expanded use of cogeneration to produce electricity and steam during the upgrading stages of oil sands production, particularly for in-situ production. Bio-upgrading technology in development that includes the use of microbes to remove sulfur compounds and impurities. Use of co-products (e.g., petroleum coke) as replacement fuels for coal-fired power generation.
	Storage	Carbon capture and storage (CCS) technologies to store CO_2 produced from point sources.
	Vented Emissions	Vapor recovery units where possible, flares otherwise.
Moderate	Land-Use Changes	Reclamation.
	Capital Equipment and Infrastructure	Energy-efficiency measures.
Small	Transportation	Energy-efficiency measures.
	Fugitive Emissions	Leak detection and repair.

Source: CRS, from studies outlined in Table 1.

Notes: Significant = greater than approximately 3% change in WTW emissions. Moderate = approximately 1%– 3% change in WTW emissions. Small = less than approximately 1% change in WTW emissions.

End Notes

[1] The resource has been referred to by several terms, including oil sands, tar sands, and, most technically, bituminous sands. Because of its widespread use in academic literature, the term "oil sands" is used in this report.

[2] For more information on oil sands resources, see CRS Report RL34258, *North American Oil Sands: History of Development, Prospects for the Future*, by Marc Humphries.

[3] For a full analysis of TransCanada's Keystone XL Pipeline project, see CRS Report R41668, *Keystone XL Pipeline Project: Key Issues*, by Paul W. Parfomak et al., and CRS Report R42124, *Proposed Keystone XL Pipeline: Legal Issues*, by Adam Vann et al.

[4] A list of studies surveyed in this report can be found in Table 1; an account of the finding can be found in Table 2.

[5] The full report by the State Department's contractor, IFC International LLC, is found in U.S. Department of State, *Keystone XL Project, Final Environmental Impact Statement, Appendix V*, "Life-Cycle Greenhouse Gas Emissions of Petroleum Products from WCSB Oil Sands Crudes Compared with Reference Crudes," July 13, 2011, at http://www.keystonepipeline-xl.state.gov/.

[6] Several of the studies, however, question this finding, and in particular, whether the production of Canadian oil sands crude would be economically viable if not exported through pipelines to the United States. See, for example, Natural Resources Defense Council, "Say No to Tar Sands Pipeline," March 2011, at http://www.nrdc.org/land/files/ TarSandsPipeline4pgr.pdf.

[7] Greenhouse gases include carbon dioxide (CO2), methane (CH4), nitrous oxide (N2O), hydrofluorocarbons (HFCs), perfluorocarbons (PFCs), and sulfur hexafluoride (SF6), among many others. In order to compare and aggregate different greenhouse gases, various techniques have been developed to index the effect each greenhouse gas has to that of carbon dioxide, where the effect of CO2 equals one. When the various gases are indexed and aggregated, their combined quantity is described as the CO2-equivalent. In other words, the CO2-equivalent quantity would have the same effect on, say, radiative forcing of the climate, as the same quantity of CO2.

[8] A study's choice of boundaries is responsible for many of the vastly differing values for GHG emissions that are currently being reported in published studies of the Canadian oil sands relative to other reference crudes. For

example, when expressed on a WTT basis rather than on a WTW basis, GHG emissions from Canadian oil sands crudes may show values that are significantly higher than reference crudes due to the technical omission of combustion from the calculation (see the reported findings in subsequent sections for examples).

[9] For more technical information on bitumen, see CRS Report RL34258, *North American Oil Sands: History of Development, Prospects for the Future*, by Marc Humphries; and, for example, National Petroleum Council, *Heavy Oil, Topic Paper #22,* July 18, 2007, at http://www.npc.org/study_topic_papers/22-ttg-heavy-oil.pdf.

[10] Predictions range from 50% in IHS CERA, *Oil Sands, Greenhouse Gases, and U.S. Oil Supply: Getting the Numbers Right*, IHS Cambridge Energy Research Associates, Inc., 2010, to 40% in Canadian Association of Petroleum Producers, "Crude Oil Forecast," June 2011.

[11] Global-warming potential (GWP) is a relative measure of how much heat a greenhouse gas traps in the atmosphere. It compares the amount of heat trapped by a certain mass of the gas in question to the amount of heat trapped by a similar mass of carbon dioxide. A GWP is calculated over a specific time interval, commonly 20, 100, or 500 years. All data included in this report use a 100-year time interval.

[12] LCAs often characterize emissions into primary and secondary flows. Primary flows are associated with the various stages in the hydrocarbon life cycle, from extraction of the resource to the combustion of the final refined fuel. Primary flows are generally well understood and included in most LCAs. Secondary flows are associated with activities not directly related to the conversion of the hydrocarbon resource into useful product (e.g., local and indirect land-use changes, construction emissions, etc.). Because these flows are outside the primary operations, they are often characterized differently across studies or excluded from LCAs altogether.

[13] For a discussion of the role and effects of greenhouse gases in climate change, see CRS Report RL34266, *Climate Change: Science Highlights*, by Jane A. Leggett.

[14] The heating value or energy value of gasoline is the amount of heat released during the combustion of a specified amount of it. The quantity known as higher heating value (HHV) is determined by bringing all the products of combustion back to the original pre-combustion temperature, and in particular condensing any vapor produced. The quantity known as lower heating value (LHV) assumes that the latent heat of vaporization of water in the fuel and the reaction products is not recovered. LHV is useful in

comparing transportation fuels because condensation of the combustion products is not practical.

[15] This baseline value is based on NETL 2008. It assesses the emissions from a weighted average of imported transportation fuels to the United States in 2005. It does not, however, include emissions from some of the most carbon-intensive imported crude oils (e.g., Venezuelan Heavy) due to modeling uncertainties. This baseline is consistent with the definitions for "baseline life-cycle greenhouse gas emissions" as used in the Energy Independence and Security Act (EISA) of 2007 and the U.S. Renewable Fuel Standards Program of 2010.

[16] Weighted average computations refer to the assumed mix of crude types and production processes that make up the bulk of the final transported product (e.g., mining versus in-situ, SCO versus dilbit). The assumptions are based on reported industry practices, and are reported differently in each study.

[17] As described in IFC 2011, diluting raw bitumen with light hydrocarbons creates what is referred to as a "dumbbell" blend, since it contains high fractions of both the heavy residuum and the light ends, with relatively low fractions of hydrocarbons in the middle that can be easily refined into premium fuel products. As a result, producing one barrel of premium fuel products (i.e., gasoline, diesel, and jet fuel) requires more dilbit input and produces more light ends and petroleum coke than refining one barrel of premium fuel products from other crudes and SCO. This results in additional energy use and GHG emissions from refining the dilbit, and producing, distributing, and combusting the light- and heavy-end co-products.

[18] Environment Canada, National Inventory Report: 1990-2008 Greenhouse Gas Sources and Sinks in Canada, 2010.

[19] Yeh, S., et al., "Land Use Greenhouse Gas Emissions from Conventional Oil Production and Oil Sands," *Environ. Sci. Technol.*, 2010, 44 (22), pp. 8766–8772.

[20] Bergerson, J. & Keith, D., Life Cycle Assessment of Oil Sands Technologies, Paper No. 11 of the Alberta Energy Futures Project, University of Calgary, 2006; J. Bergerson, The Impact of LCFS on Oil Sands Development: Hybrid LCA Methods, Presentation at the InLCA/LCM Conference, October 2, 2007, University of Calgary.

[21] See, for example, Rooney, R., et al., *Oil Sands Mining and Reclamation Cause Massive Loss of Peatland and Stored Carbon*, PNAS, at http://www.pnas.org/cgi/doi/10.1073/pnas. 1117693108.

[22] NETL 2009 assumes the production of these specific reference crudes could be affected most by an increase in Canadian oil sands production. See next section "Design Factors and Input Assumptions for Reference Crudes Assessments."

[23] Many factors—from economics, to geopolitics, to trade issues—would influence the balance of global petroleum production. An analysis of how incremental production of Canadian crudes would affect the production levels of other global crudes, and which of those crudes would be backed out of U.S. refineries and/or global production, is beyond the scope of this report. For more detail on global oil markets, see CRS Report R41765, *U.S. Oil Imports: Context and Considerations*, by Neelesh Nerurkar.

[24] DOS, Final EIS, 3.14.52-56.

[25] See 2012 Draft U.S. Greenhouse Gas Inventory Report, Table-ES-2, at http://epa.gov/climate change/emissions/downloads12/Executive%20Summary.pdf.

[26] Final EIS, 3.14-55.

In: Canadian Oil Sands, Greenhouse Gases ... ISBN: 978-1-62257-402-5
Editors: T. J. Carlson and S. Harper

Chapter 2

NORTH AMERICAN OIL SANDS: HISTORY OF DEVELOPMENT, PROSPECTS FOR THE FUTURE*

Marc Humphries

SUMMARY

When it comes to future reliable oil supplies, Canada's oil sands will likely account for a greater share of U.S. oil imports. Oil sands account for about 46% of Canada's total oil production and oil sands production is increasing as conventional oil production declines. Since 2004, when a substantial portion of Canada's oil sands were deemed economic, Canada, with about 175 billion barrels of proved oil sands reserves, has ranked second behind Saudi Arabia in oil reserves. Canadian crude oil exports were about 1.82 million barrels per day (mbd) in 2006, of which 1.8 mbd or 99% went to the United States. Canadian crude oil accounts for about 18% of U.S. net imports and about 12% of all U.S. crude oil supply.

Oil sands, a mixture of sand, bitumen (a heavy crude that does not flow naturally), and water, can be mined or the oil can be extracted in-situ using thermal recovery techniques. Typically, oil sands contain about 75% inorganic matter, 10% bitumen, 10% silt and clay, and 5% water. Oil sand is sold in two forms: (1) as a raw bitumen that must be blended

---

* This is an edited, reformatted and augmented version of Congressional Research Service., Publication No. RL34258, dates Jan 2008.

with a diluent for transport and (2) as a synthetic crude oil (SCO) after being upgraded to constitute a light crude. Bitumen is a thick tar-like substance that must be upgraded by adding hydrogen or removing some of the carbon.

Exploitation of oil sands in Canada began in 1967, after decades of research and development that began in the early 1900s. The Alberta Research Council (ARC), established by the provincial government in 1921, supported early research on separating bitumen from the sand and other materials. Demonstration projects continued through the 1940s and 1950s. The Great Canadian Oil Sands company (GCOS), established by U.S.-based Sunoco, later renamed Suncor, began commercial production in 1967 at 12,000 barrels per day.

The U.S. experience with oil sands has been much different. The U.S. government collaborated with several major oil companies as early as the 1930s to demonstrate mining of and in-situ production from U.S. oil sand deposits. However, a number of obstacles, including the remote and difficult topography, scattered deposits, and lack of water, have resulted in an uneconomic oil resource base. Only modest amounts are being produced in Utah and California. U.S. oil sands would likely require significant R&D and capital investment over many years to be commercially viable. An issue for Congress might be the level of R&D investment in oil sands over the long term.

As oil sands production in Canada is predicted to increase to 2.8 million barrels per day by 2015, environmental issues are a cause for concern. Air quality, land use, and water availability are all impacted. Socio-economic issues such as housing, skilled labor, traffic, and aboriginal concerns may also become a constraint on growth. Additionally, a royalty regime favorable to the industry has recently been modified to increase revenue to the Alberta government. However, despite these issues and potential constraints, investment in Canadian oil sands will likely continue to be an energy supply strategy for the major oil companies.

INTRODUCTION

Current world oil reserves are estimated at 1.292 trillion barrels. The Middle East accounts for 58% of world oil reserves, and the Organization of Petroleum Exporting Countries (OPEC) accounts for 70%. The Middle East also leads in reserve growth and undiscovered potential, according to the Energy Information Administration (EIA).[1]

The United States' total oil reserves are estimated at 22.7 billion barrels, a scant 1.8% of the world's total (see Appendix A). U.S. crude oil production is expected to fall from 5.4 million barrels per day (mbd) in 2004 to 4.6 mbd in 2030, while demand edges up at just over 1% annually. Net imports of petroleum are estimated by the EIA to increase from 12.1 mbd (58% of U.S. consumption) to 17.2 mbd (62% of U.S. consumption) over the same time period.[2]

When it comes to future reliable oil supplies, Canadian oil sands will likely account for a larger share of U.S. oil imports. Oil sands account for about 46% of Canada's total oil production, and oil sand production is increasing as conventional oil production declines. Since 2004, when a substantial portion of Canada's oil sands were deemed economic, Canada has been ranked second behind Saudi Arabia in oil reserves. Canadian crude oil exports were about 1.82 million barrels per day in 2006, of which 1.8 mbd or 99% went to the United States. Canadian crude oil accounts for about 18% of U.S. net imports and about 12% of all U.S. crude oil supply.

An infrastructure to produce oil, upgrade, refine, and transport it from Canadian oil sand reserves to the United States is already in place. Oil sands production is expected to rise from its current level of 1.2 (mbd) to 2.8 mbd by 2015. However, infrastructure expansions and skilled labor are necessary to significantly increase the flow of oil from Canada. For example, many refineries are optimized to refine only specific types of crude oil and may not process bitumen from oil sands. One issue likely to be contentious is the regulatory permitting of any new refinery capacity because of environmental concerns such as water pollution and emissions of greenhouse gases.

Challenges such as higher energy costs, infrastructure requirements, and the environment, may slow the growth of the industry. For example, high capital and energy input costs have made some projects less economically viable despite recent high oil prices. Canada ratified the Kyoto Protocol in 2002, which bound Canada to reducing its greenhouse gas (GHG) emissions significantly by 2012 but according to the government of Canada they will not meet their Kyoto air emission goals by 2012. The Pembina Institute reports that the oil sands industry accounts for the largest share of GHG emissions growth in Canada.[3]

Major U.S. oil companies (Sunoco, Exxon/Mobil, Conoco Phillips, and Chevron) continue to make significant financial commitments to develop Canada's oil sand resources. Taken together, these companies have already committed several billion dollars for oil sands, with some projects already operating, and others still in the planning stages. Many of these same firms, with the U.S. government, did a considerable amount of exploration and development on "tar sands" in the United States, conducting several pilot projects. These U.S. pilot projects did not prove to be commercially viable for oil production and have since been abandoned. Because of the disappointing results in the United States and the expansive reserves in Canada, the technical expertise and financial resources for oil sands development has shifted almost exclusively to Canada and are likely to stay in Canada for the foreseeable future. However, with current oil prices above $60 per barrel and the possibility of sustained high prices, some oil sand experts want to re-evaluate the commercial prospects of U.S. oil sands, particularly in Utah.

This report examines the oil sands resource base in the world, the history of oil sands development in the United States and Canada, oil sand production, technology, development, and production costs, and the environmental and social impacts. The role of government — including direct financial support, and tax and royalty incentives — is also assessed.

WORLD OIL SANDS RESERVES AND RESOURCES[4]

Over 80% of the earth's technically recoverable natural bitumen (oil sands) lies in North America, according to the U.S. Geological Survey (USGS) (see Appendix B). Canadian oil sands account for about 14% of world oil reserves and about 11% of the world's technically recoverable oil resources.

What Are Oil Sands?

Oil sands (also called tar sands) are mixtures of organic matter, quartz sand, bitumen, and water that can either be mined or extracted in-situ[5] using thermal recovery techniques. Typically, oil sands contain about 75% inorganic matter, 10% bitumen, 10% silt and clay, and 5% water.[6] Bitumen is a heavy crude that does not flow naturally because of its low API[7] (less than 10 degrees) and high sulfur content. The bitumen has high density, high viscosity, and high metal concentration. There is also a high carbon-to-hydrogen molecule count (i.e. oil sands are low in hydrogen). This thick, black, tar-like substance must be upgraded with an injection of hydrogen or by the removal of some of the carbon before it can be processed.

Oil sand products are sold in two forms: (1) as a raw bitumen that must be blended with a diluent[8] (becoming a bit-blend) for transport and (2) as a synthetic crude oil (SCO) after being upgraded to constitute a light crude. The diluent used for blending is less viscous and often a by-product of natural gas, e.g., a natural gas condensate. The specifications for the bit blend (heavy oil) are 21.5 API and a 3.3% sulfur content and for the SCO (light oil) are 36 API and a 0.015% sulfur content.[9]

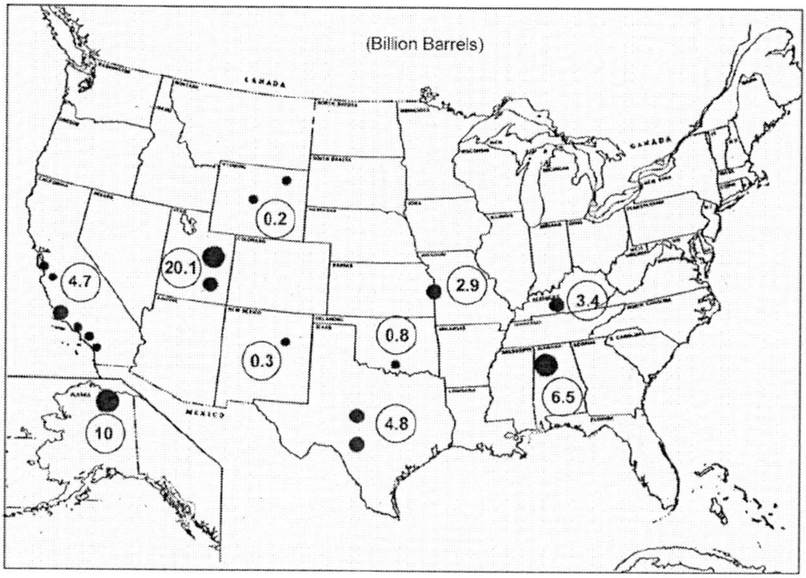

Source: *Major Tar Sand and Heavy Oil Deposits of the United States,* Interstate Oil Compact Commission, 1984, p. 2.

Figure 1. Tar (Oil) Sand Deposits of the United States

U.S. Oil Sand Resources

The USGS, in collaboration with the U.S. Bureau of Mines, concluded in a 1984 study that 53.7 billion barrels (21.6 billion measured plus 32.1 billion speculative) of oil sands could be identified in the United States. An estimated 11 billion barrels of those oil sands could be recoverable. Thirty-three major deposits each contain an estimated 100 million barrels or more. Fifteen percent were considered mineable and 85% would require in-situ production. Some of the largest measured U.S. oil sand deposits exist in Utah and Texas. There are smaller deposits located in Kentucky, Alabama, and California. Most of the deposits are scattered throughout the various states listed above. As of the 1980s, none of these deposits were economically recoverable for oil supply. They are still not classified as reserves (see Figure 1).

Canadian Oil Sand Resources

Canadian oil sand resources are located almost entirely in the province of Alberta. The Alberta Energy and Utility Board (AEUB) estimates that there are 1.6 trillion barrels of oil sands in place, of which 11% are recoverable (175 billion barrels) under current economic conditions (see Table 1). Mineable reserves at the surface account for 35 billion barrels (20%) and in-situ reserves at 141 billion barrels (80%). The AEUB estimates that the ultimate amount to be discovered (ultimate volume-in place) is 2.5 trillion barrels: about 2.4 trillion in-situ and 140 billion surface-mineable. Of this ultimate discovered amount, about 314 billion barrels are expected to be recovered (175 billion barrels in reserves now and another 143 billion barrels anticipated. See Table 1).[10] However, EIA estimates only 45.1 billion barrels (reserve growth and undiscovered potential) to be added to Canada's reserve base by 2025.[11]

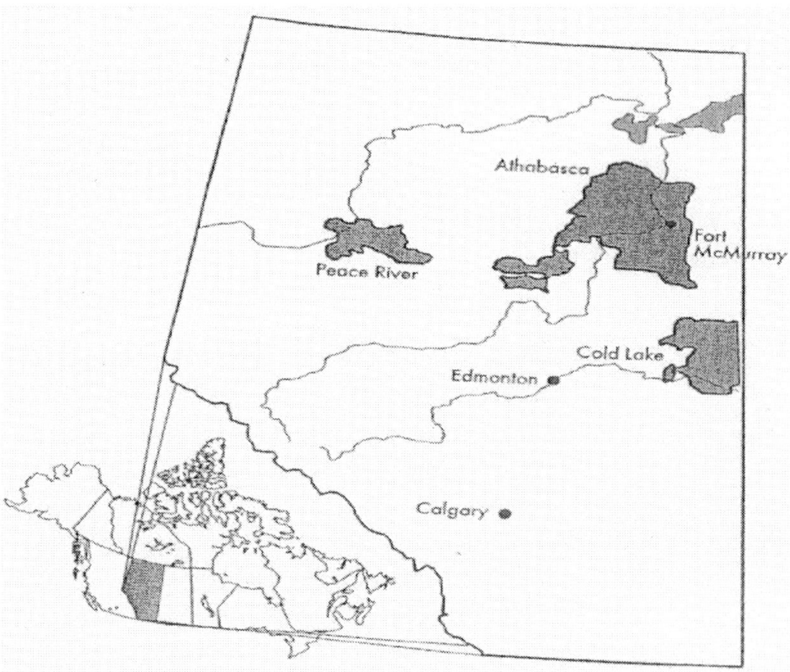

Source: National Energy Board, Alberta, Canada.

Figure 2. Oil Sands Areas in Alberta, Canada

Oil sands occur primarily in three areas of Alberta: Peace River, Athabasca, and Cold Lake (see Figure 2 below). Current production is 1.1 million barrels per day and is expected to reach 2.0 mbd by 2010 and 3.0 mbd by 2015.[12] According to the International Energy Agency (IEA), Canada's oil sands production could exceed 5.0 mbd by 2033 but would require at least $90 billion in investment.[13]

As a result of recent high oil prices, 44 new oil sands projects are planned for Alberta between 2004 and 2012, 26 in-situ and 18 surface-mining.[14] If all projects were to go forward, an estimated C$60 billion would be required for construction. Several of the projects are expansions of current operations. The National Energy Board (NEB) projects as much as C$81.6 billion being spent between 2006 and 2016.[15] Eighty-two percent of the projected investment — expected to peak in 2008 — is directed towards the Fort McMurray/Woods Buffalo Region of Alberta. A total of C$29 billion was spent on oil sands development between 1996 and 2004.[16]

Table 1. Canada's Bitumen Resources

Billion Barrels	Ultimate Volume in Place	Initial Volume in Place	Ultimate Recoverable Volume	Initial Established Reserves	Cumulative Production	Remaining Established Reserves
Mineable						
Athabasca	138.0	113.0	69.0	35.0	2.5	32.7
In Situ						
Athabasca	N/A	1,188.0	N/A	N/A	N/A	N/A
Cold Lake	N/A	201.0	N/A	N/A	N/A	N/A
Peace River	N/A	129.0	N/A	N/A	N/A	N/A
Subtotal	2,378.0	1,518.0	245.0	142.8	1.26	141.5
Total	**2,516.0**	**1,631.0**	**314.0**	**177.8**	**3.76**	**174.2**

Source: Alberta Energy Utility Board.

History of Development

Role of Industry and Government

U.S. Oil Sands. Interest in U.S. oil sand deposits dates back to the 1930s. Throughout the 1960s and 1970s, 52 pilot projects involving mining and in-situ techniques were supported by the U.S. government in collaboration with major oil companies such as Conoco, Phillips Petroleum, Gulf Oil, Mobil,

Exxon, Chevron, and Shell. Several steam-assisted technologies were being explored for in-situ production. These sources have had little economic potential as oil supply. The Energy Policy Act of 2005 (P.L. 109-58), however, established a public lands leasing program for oil sands and oil shale[17] R&D.

Based on the Canadian experience with oil sands production, it was established that commercial success in mining oil sands is a function of the ratio of overburden to oil sand thickness.[18] This ratio should not exceed one. In other words, the thickness of the overlying rock should not be greater than the thickness of the sand deposit. It was estimated by the USGS that only about 15% of the U.S. resource base has a ratio of one or less.

Major development obstacles to the U.S. oil sands resource base include remote and difficult topography, scattered deposits, and the lack of water for in-situ production (steam recovery and hot water separation) or undeveloped technology to extract oil from U.S. "hydrocarbon-wetted" deposits.[19] The Canadian technology may not be suited for many U.S. deposits. In Texas, deposits were considered by Conoco Oil to be too viscous to produce in-situ. A Bureau of Mines experiment with oil sands production in Kentucky proved to be commercially infeasible. In Utah, there were attempts at commercial production over the past three decades by several oil companies but projects were considered uneconomic and abandoned. As of 2004, some oil sands were being quarried on Utah state lands for asphalt used in road construction, and a small amount of production is taking place in California.[20] "Since the 1980s there has been little production for road material and no government funding of oil sands R&D," according to an official at the Department of the Interior.[21]

A 2006 conference on oil sands held at the University of Utah indicated renewed interest in U.S. oil sands but reiterated the development challenges mentioned above. Speakers also pointed out new technologies on the horizon that are being tested in Utah.[22] Conference organizers concurred that long-term research and development funding and huge capital development costs would be needed to demonstrate any commercial potential of U.S. oil sand deposits. A recent report[23] on U.S. unconventional fuels (an interagency and multistate collaboration) makes a number of general recommendations (for the development of oil sands and other unconventional fuels), which include economic incentives, establishing a regulatory framework, technology R&D, and an infrastructure plan. A recommendation specific to oil sands calls for closer U.S. collaboration with the government of Alberta to better understand Canadian oil sands development over the last 100 years. The report's task force estimates that based on a "measured" or "accelerated" development pace scenario,[24] U.S. oil sand production could reach 340,000-352,000 barrels per day by 2025.[25]

Canadian Oil Sands. Canada began producing its oil sands in 1967 after decades of research and development that began in the early 1900s. Wells were drilled between 1906 and 1917 in anticipation of finding major conventional oil deposits. The area around Fort McMurray, Alberta, was mapped for bituminous sand exposures in 1913 by Canada's Federal Department of Mines. By 1919, the Scientific and Industrial Research Council of Alberta (SIRCA), predecessor to the Alberta Research Council (ARC),[26] became interested in oil sands development. One of its newly recruited scientists, Dr. Karl Clark, began his pioneering work on a hot-water flotation process for separating the bitumen from the sand. In this separation process, the mined oil sand is mixed with water and a sodium hydroxide base and rotated[27] in a horizontal drum at 80 degrees centigrade. Dr. Clark's efforts led to a pilot plant in 1923 and a patented process by 1929. He continued to improve the process through several experimental extraction facilities through the 1940s.

The technical feasibility was demonstrated in 1949 and 1950 at a facility in Bitumont, Alberta, located on the Athabasca River near Fort McMurray. The technology being tested was largely adopted by the early producers of oil sands —Great Canadian Oil Sands (GCOS), Ltd., and Syncrude. Sunoco established GCOS, Ltd., in 1952 and then invested $250 million in its oil sands project. Another major player in the oil sands business in Canada was Cities Services, based in Louisiana. Cities Services purchased a controlling interest in the Bitumont plant in 1958, then in 1964, along with Imperial Oil, Atlantic Richfield (ARCO), and Royalite Oil, formed the Syncrude consortium.[28]

The ARC continued its involvement with oil sands R&D throughout the 1950s and 1960s. Several pilot projects were established during that period. Suncor [29] began construction of the first commercial oil sands production/separation facility in 1964 and began production in 1967, using the hot water extraction method developed and tested by ARC. In 1967, Suncor began to produce oil sands at a rate of 12,000 barrels per day.

Just a year later, in 1968, the government of Alberta deferred an application by Syncrude Canada for a $200 million, 80,000 barrel oil sands facility. Eventually, in 1978, the Energy Resources Conservation Board of Alberta approved Syncrude's proposal to build a $1 billion plant that would produce up to 129,000 barrels per day.

However, ARCO, which represented 30% of the project, pulled out of the consortium as costs of the plant climbed toward $2 billion. At that point (1978) the federal and provincial governments joined in. The federal government purchased a 15% share, Alberta a 10% share, and Ontario 5%, making up the 30% deficit. At the time, the Canadian government was promoting the goal of energy self-sufficiency, and the Alberta government agreed to a 50/50 profit-sharing arrangement instead of normal royalties for Syncrude.[30]

The Alberta Energy Company[31] purchased 20% of Syncrude and then sold 10% of its share to Petrofina Canada, Ltd., and Hudson Bay Oil and Gas, Ltd.[32] The consortium grew from four to nine owners. From 1983 to 1988 Syncrude spent $1.6 billion to boost production to 50 million barrels per year. In 1984, the government of Alberta agreed to a new royalty structure for oil sands producers coinciding with Syncrude's capital expansion plans. In 1985, the Alberta government announced that existing oil sands operations and new plants would not be taxed on revenues, and the petroleum gas revenue tax would be phased out. During the same time-frame, Syncrude's cash operating costs were just under $18 per barrel with total costs over $20 per barrel,[33] while the market price of oil fluctuated under $20 per barrel.

Because of huge capital requirements, oil sands producers lobbied for continued royalty relief and thought the government should "defer tax and royalty revenues until project expansions were completed."[34] In 1994, the National Oil Sands Task Force (an industry/government group) was created, and the Canadian Oil Sands Network for R&D (CONRAD) agreed to spend $105 million annually to boost production and trim costs. Costs continued to fall ($15.39/bbl in 1992[35] to under $14/ bbl in 1994[36]) as Syncrude ownership continued to change. In 1996, the National Oil Sands Task Force recommended a package of royalty and tax terms to ensure consistent and equal treatment of projects, because oil sand projects previously were treated on a project-by-project basis. The implementation of favorable royalty treatment is discussed below.

The ARC has had a successful partnership with the private sector in oil sands research and development. As a result of favorable royalty and tax terms and Alberta's $700 million R&D investment in oil sands extraction (from 1976-2001), the private sector has invested billions of dollars of development capital in oil sand projects.[37] Syncrude has said that "partnering with ARC gave us the ability to explore a potentially valuable technology."[38]

Oil Sands Production Process

Oil sands production measured only 1.3% of total world crude oil production in 2005. By 2025 it may reach 4.1% of total world production. But more importantly, it may mean U.S. access to extensive North American oil reserves and increased energy security.

Oil sands are either surface-mined or produced in-situ. Mining works best for deposits with overburden less than 75 meters thick. Mining requires a hydraulic or electric shovel that loads the sand into 400-ton trucks, which carry the material to a crusher to be mixed into a slurry. Using pumps and pipelines, the slurry is "hydro transported" to an extraction facility to extract bitumen (see Figure 3). This process recovers about 90% of the bitumen.[39]

In 2005, mining accounted for about 52% of Alberta's oil sand production (572,000 b/d); in-situ accounted for about 48% (528,000 b/d), one-third of which was produced using the Cold Production method in which oil sands are light enough to flow without heat. The in-situ approach, which was put into commercial production in 1985,[40] is estimated to grow to 926,000 barrels per day by 2012. Currently, the largest production projects are in the Fort McMurray area operated by Syncrude and Suncor (see Table 4 for leading producers of oil sands).

Extraction Process. The extraction process separates the bitumen from oil sands using warm water (75 degrees Fahrenheit) and chemicals. Extracting the oil from the sand after it is slurried consists of two main steps. First is the separation of bitumen in a primary separation vessel. Second, the material is sent to the froth tank for diluted froth treatment to recover the bitumen and reject the residual water and solids. The bitumen is treated either with a naphtha solvent or a paraffinic solvent to cause the solids to easily settle. The newer paraffinic treatment results in a cleaner product.[41] This cleaner bitumen is pipeline quality and more easily blended with refinery feedstock. After processing, the oil is sold as raw bitumen or upgraded and sold as SCO.

Production Technology. For in-situ thermal recovery, wells are drilled, then steam is injected to heat the bitumen so it flows like conventional oil. In-situ production involves using various techniques.

One technique is the Cyclic Steam Stimulator (CSS), also known as "huff and puff." CSS is the most widely used in-situ technology. In this process, steam is added to the oil sands via vertical wells, and the liquefied bitumen is pumped to the surface using the same well.

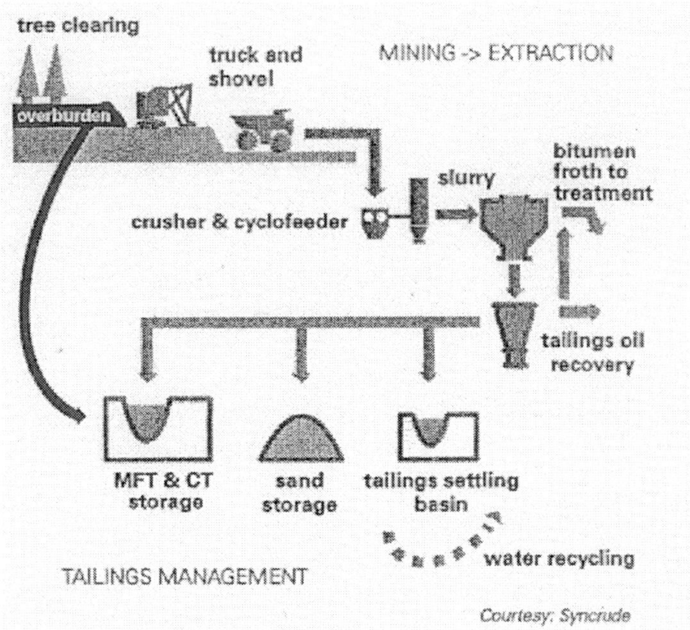

Source: *Oil Sands Technology Roadmap*, Alberta Chamber of Resources, January 2004, p. 21.

Figure 3. Major Mining Process Steps

Table 2. Leading Oil Sands Producers (barrels per day)

Project Owner	Type of Project	2002	2003	2006 1st Quarter	Planned Production Targets
Suncor	Mining	206,000	217,000	264,400	410,000
Syncrude	Mining	230,000	212,000	205,000	560,00
Athabasca Oil Sands (Shell, Chevron, and Marathon Oil)[a]	Mining	N/A	130,000	77,400	525,000
Imperial Oil	In-situ	112,000	130,000	150,000	180,000
CNRL	In-situ	N/A	35,000	122,000	500,000
Petro Canada	In-situ	4,500	16,000	21,000 (2005)	100,000
EnCana	In-situ	N/A	5,300	36,000	250,000

Source: *Oil Sands Industry Update*, Alberta Economic Development, 2004 and 2006.

a. Marathon Oil Corp. acquired Western Oil Sands, Inc. on October 18, 2007.

But a relatively new technology — steam-assisted gravity drainage (SAGD) —has demonstrated that its operations can recover as much as 70% of the bitumen in-place. Using SAGD, steam is added to the oil sands using a horizontal well, then the liquefied bitumen is pumped simultaneously using another horizontal well located below the steam injection well (see Figure 4). The SAGD process has a recovery advantage over the CSS process, which only recovers 25%-30% of the natural bitumen. Also, the lower steam to oil ratio (the measurement of the volume of steam required to extract the bitumen) of SAGD results in a more efficient process that uses less natural gas.[42] SAGD operations are limited to thick, clean sand reservoirs, but it is reported by the industry that most of the new in-situ projects will use SAGD technology.[43] A number of enhanced SAGD methods are being tested by the Alberta Research Council. They could lead to increased recovery rates, greater efficiency, and reduced water requirements.

The emerging Vapor Extraction Process (VAPEX) technology operates similarly to SAGD. But instead of steam, ethane, butane, or propane is injected into the reservoir to mobilize the hydrocarbons towards the production well. This process eliminates the cost of steam generators and natural gas. This method requires no water and processing or recycling and is 25% lower in capital costs than the SAGD process. Operating costs are half that of the SAGD process.[44]

A fourth technique is cold production, suitable for oil sands lighter than those recovered using thermal assisted methods or mining. This process involves the co-production of sand with the bitumen and allows the oil sands to flow to the well bore without heat. Imperial Oil uses this process at its Cold Lake site. Oil sand produced using in-situ techniques is sold as natural bitumen blended with a diluent for pipeline transport.

The overall result of technology R&D has been to reduce operating costs from over $20/barrel in the early 1970s to $8-12/barrel in 2000. While technology improvements helped reduce some costs since 2000, total costs have risen significantly as discussed below, because of rising capital and energy costs.[45]

Upgrading.[46] Upgrading the bitumen uses the process of coking for carbon removal or hydro-cracking for hydrogen addition (see Figure 5).

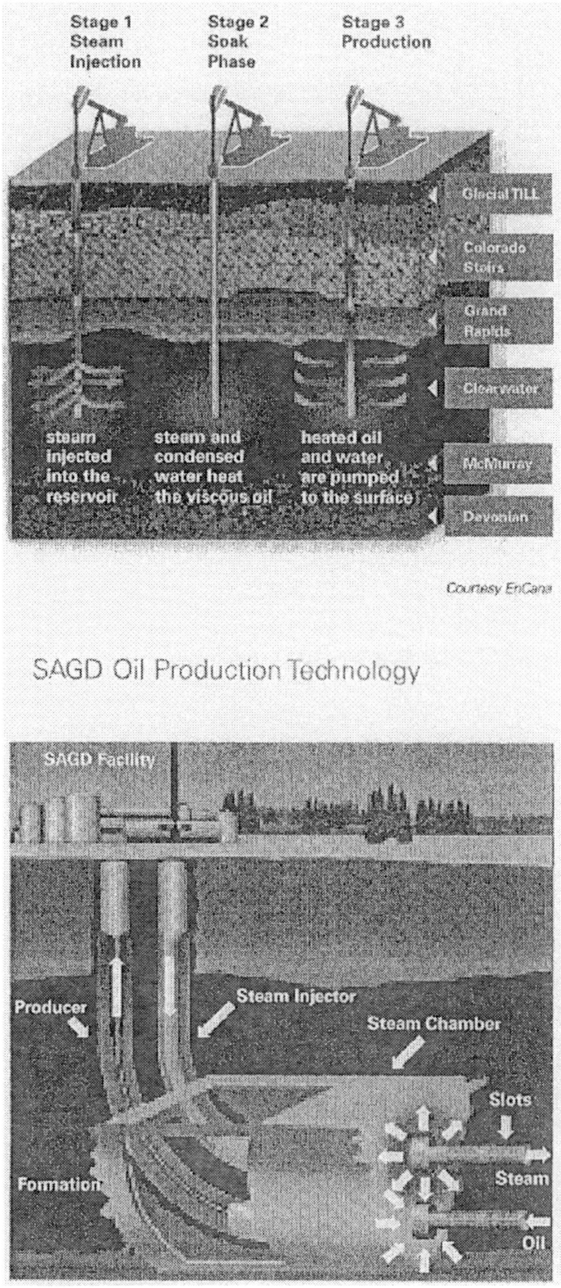

Source: *Oil Sands Technology Roadmap*, p. 28.

Figure 4. In-SITU Recovery

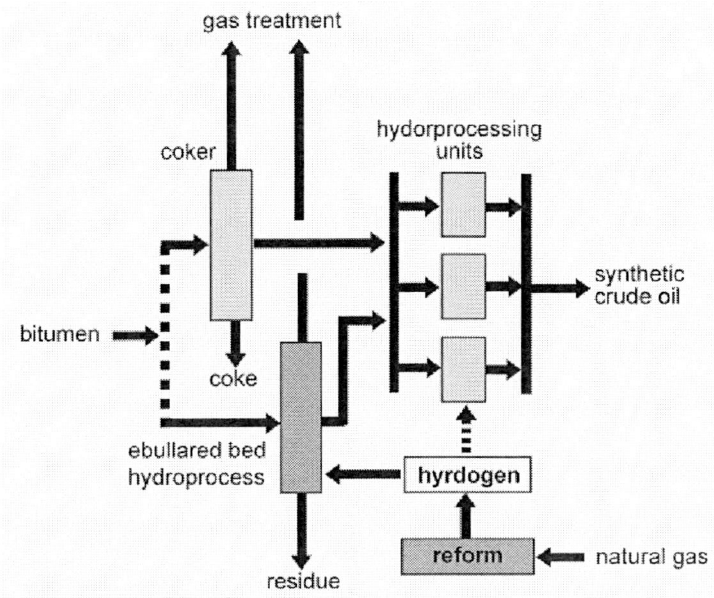

Source: *Oil Sands Technology Roadmap*, p. 41.

Figure 5. Upgrading to SCO

Coking is a common carbon removal technique that "cracks" the bitumen using heat and catalysts, producing light oils, natural gas, and coke (a solid carbon byproduct). The coking process is highly aromatic and produces a low quality product. The product must be converted in a refinery to a lighter gas and distillate. Hydrocracking also cracks the oil into light oils but produces no coke byproduct. Hydrocracking requires natural gas for conversion to hydrogen. Hydrocracking, used often in Canada, better handles the aromatics. The resulting SCO has zero residues which help keep its market value high, equivalent to light crude.

Partial upgrading raises the API of the bitumen to 20-25 degrees for pipeline quality crude. A full upgrade would raise the API to between 30-43 degrees — closer to conventional crude. An integrated mining operation includes mining and upgrading. Many of the mining operations have an on-site upgrading facility, including those of Suncor and Syncrude. Suncor uses the coking process for upgrading, while Syncrude uses both coking and hydrocracking and Shell uses hydrocracking. (For the complete oil sands processing chain, see Figure 6.)

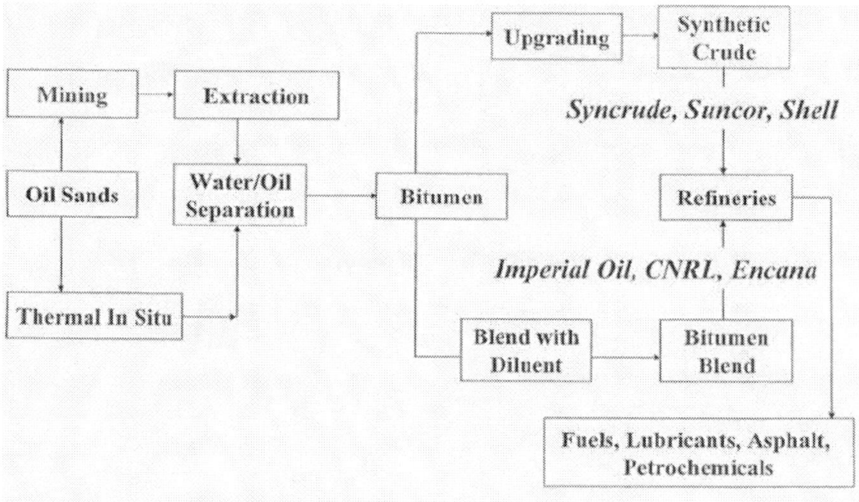

Source: *Overview of Canada's Oil Sands*, TD Securities, p. 15.

Figure 6. Oil Sands Processing Chain

A major trend among both mining and in situ producers is to integrate the upgrading with the refinery to cut costs; e.g., linking SAGD production with current refinery capabilities. Long-term processing success of oil sands will depend on how well this integration takes place and how well the industry addresses the following issues:

- cost overruns,
- cost effective upgrading, reducing highly aromatic, high-sulfur SCO, and
- dependence on and price of natural gas for hydrogen production (originally used because of its low price but now considered by some to be too expensive).

The wide heavy-oil/light-oil price differential has been an incentive to increase upgrading. The price for heavy crude was as low as $12 per barrel in early 2006 and its market is limited by refineries that can process it and by its end use as asphalt. In its June 2006 report, the NEB describes numerous proposals for building upgraders.[47]

Cost overruns for the integrated mining projects or expansions, sometimes as much as 50% or more of the original estimates, have been a huge problem for the industry. The main reasons cited by the COS report are poor management, lack of skilled workers, project size, and engineering issues.

Cost of Development and Production. Operating and total supply costs have come down significantly since the 1970s. Early supply costs were near C$35 per barrel (in 1970s dollars). Reductions came as a result of two major innovations in the production process. First, power shovels and energy efficient trucks replaced draglines and bucketwheel reclaimers, and second, hydrotransport replaced conveyor belts to transport oil sands to the processing plant.[48]

Operating costs include removal of overburden, mining and hydro transport, primary extraction, treatment, and tailings removal. The recovery rate, overburden volumes, cost of energy, transport distances, and infrastructure maintenance all have an impact on operating costs.

Supply costs (total costs) include the operating costs, capital costs, taxes and royalties, plus a 10% return on investment (ROI). When compared to conventional new oil production starts, an oil sands project may have operating costs over 30% higher than the world average for conventional new starts. However, its nearly nonexistent royalty and tax charge makes the total cost per barrel of energy significantly less than the conventional oil project. The NEB in its Energy Market Assessment estimated that between US$30-$35 per barrel oil is required to achieve a 10% ROI.[49]

Operating costs for mining bitumen were estimated at around C$9-$12 per barrel (C$2005) — an increase of up to C$4 per barrel since the 2004 NEB estimates. Supply cost of an integrated mining/upgrading operation is between C$36 and $40/ barrel for SCO — a dramatic increase over the C$22-$28 estimate made in 2004. These supply costs for an integrated mining/upgrading operation were expected to decline with improvements in technologies (see Table 3). However, natural gas prices rose 88% and capital costs rose 45% over the past two years.

Operating costs for SAGD in-situ production in 2005 were about C$10-$14 per barrel of bitumin, up from C$7.40 per barrel in 2004. Recovery rates are lower than with mining, at 40%-70%, and the price of energy needed for production is a much larger factor. The SAGD operations are typically phased-in over time, thus are less risky, make less of a "footprint" on the landscape than a mining operation, and require a smaller workforce. SAGD supply cost for Athabasca oil sand rose from between C$11-$17/barrel (bitumen) to C$18-$22/barrel; using the CSS recovery technique, supply costs are estimated higher at between C$20-$24/barrel, an increase from C$13-$19/barrel. Cost increases/decreases for in-situ operations are largely dependent on the quality of the reservoir and natural gas prices, but as SAGD and other new technologies (e.g. VAPEX) become more efficient, industry is expecting some cost declines. SAGD (in-situ) supply costs are less sensitive to capital costs than mining projects because the capital investment is far less.

Table 3. Estimated Operating and Supply Cost by Recovery Type (C$2005 Per Barrel at the Plant Gate)

	Crude Type	Operating Cost	Supply Cost
Cold Production - Wabasca, Seal	Bitumen	6-9	14-18
Cold Heavy Oil Production and Sand (CHOPS) - Cold Lake	Bitumen	8-10	16-19
Cyclic Stream Stimulation (CSS)	Bitumen	10-14	20-24
Steam Assisted Gravity Drainage (SAGD)	Bitumen	10-14	18-22
Mining/Extraction	Bitumen	9-12	18-20
Integrated Mining/Upgrading	SCO	18-22	36-40

Source: *Canada's Oil Sands, Opportunities and Challenges to 2015*, National Energy Board, Canada, June 2006.

Note: Supply costs for the first five technologies do not include the coat of upgrading bitumen to SCO.

Natural gas is a major input and cost for mining, upgrading, and in situ recovery: Mining requires natural gas to generate heat for the hot water extraction process, upgraders need it for heat and steam, and in situ producers use natural gas to produce steam which is injected underground to induce the flow of bitumen. Natural gas accounts for 15% of the operating costs in mining operations compared to 60% of operating costs in SAGD in-situ production. The major cost for thermal in-situ projects (SAGD, CSS) is for the natural gas that powers the steam-producing generators. For SAGD projects, 1 thousand cubic feet is needed per barrel of bitumen. Reducing the steam-to-oil ratio (SOR) — the quantity of steam needed per barrel of oil produced — is critical for lowering natural gas use and costs.[50] SAGD has a lower SOR than CSS projects but cannot be used for all oil sand in-situ production. However, most new in-situ projects will use SAGD.

Canadian oil sand producers continue to evaluate energy options that could reduce or replace the need for natural gas. Those options include, among other things, the use of gasification technology, cogeneration, coal, and nuclear power.

Tax and Royalty on Oil Sands. In 1997 the Alberta government implemented a "Generic Oil Sands Royalty Regime"[51] specific to oil sands for all new investments or expansions of current projects. Since then, oil sand producers have had to pay a 1% minimum royalty based on gross revenue until all capital costs including a rate of return are recovered. After that, the royalty is either 25% of net project revenues or 1% of the gross revenues, whichever is greater.[52] The 1% pre-payout royalty rate is in stark contrast to conventional world royalties. Net project revenues (essentially net profits before tax) include revenues after project cash costs, such as operating costs, capital, and R&D are deducted. Royalty payments may be based on the value of bitumen or SCO if the project includes an upgrader. Currently, 51% of oil sand projects (or 75% of production volume) under the Generic Royalty regime are paying the 25% royalty rate. Two major oil sands producers, Suncor and Syncrude (accounting for 49% of bitumen production) have "Crown Agreements" in place with the province that have allowed the firms to pay royalties based on the value of synthetic crude oil (SCO) production with the option to switch to paying royalties on the value of bitumen beginning as early as 2009. Royalties paid on bitumen, which is valued much lower than SCO, would result in less revenue for the government. The agreements expire in 2016.

Royalty revenues from oil sands fluctuated widely between 1997 and 2005. For example, royalties from oil sands were less than $100 million in 1999, then rose to $700 million in 2000/2001, but fell in 2002/2003 to about $200 million as production continued to rise. Royalties from oil sands rose dramatically in 2005/2006 to $1 billion, and the Government of Alberta forecasts royalties of $2.5 billion in 2006/2007 and $1.8 billion in 2007/2008.[53] Oil price fluctuations are the primary cause for such swings in royalty revenues.

The Albertan provincial government established a Royalty Review Panel in February 2007 to examine whether Alberta was receiving its fair share of royalty revenues from the energy sector and to make recommendations if changes are needed. In its September 2007 report, the panel concluded that "Albertans do not receive their fair share from energy development."[54] When the oil sands industry was ranked against other heavy oil and offshore producers such as Norway, Venezuela, Angola, United Kingdom, and the U.S. Gulf of Mexico, Alberta received the smallest government share.[55] This is, however, a difficult comparison to make because it is not among oil sand producers only and the fiscal regimes of the various producing countries is dynamic. However, based on a general analysis by T.D. Securities, typically, on average, world royalty rates could add as much as 45% to operating costs while the 1% rate may add only 3% to operating costs.[56]

The Panel recommended keeping the "pre-payout, post-payout" framework intact (see footnote 52), which would retain the 1% pre-payout royalty rate, but in the post-payout phase, firms would be required to pay a higher net revenue royalty rate of 33% plus continue to pay the 1% base royalty.

On October 25, 2007, the Alberta Government announced and published its response to the Royalty Review Panel's report.[57] It retained the "pre-payout," "post-payout" royalty framework but concluded that a sliding-scale rate structure would best achieve increasing the government's share of revenues from oil sands production. The pre-payout base rate would start at 1%, then increase for every dollar above US$55 per barrel (using the West Texas Intermediate or WTI price) reaching a maximum increase of 9% when prices are at or above $120 per barrel. In the post-payout phase, the net revenue rate will start at 25%, then rise for every dollar oil in priced above US$55 per barrel, reaching a maximum of 40% of net revenues when oil is $120 per barrel or higher. The new rate structure will take effect in 2009. The Government of Alberta has initiated negotiations with Suncor and Syncrude in an attempt to include them under the new oil sands royalty framework by 2009.

Oil sand firms pay federal and provincial income taxes and some differences exist in the tax treatment of the oil sands and conventional oil industries. Since the Provincial 1996 Income Tax Act, both mineable and in-situ oil sand deposits are classified as a mineral resource for Capital Cost Allowance (CCA) purposes which means mineral deposits receive higher cost deductions than conventional oil and gas operations (i.e. acquisition costs and intangible drilling costs).[58] The provincial government of Alberta has agreed to the 2007 federal budget proposal to eliminate the CCA deduction for oil sands. The Royalty Review Panel also supported this change in its report. The federal government of Canada, however, provided some balance by reducing the general federal corporate income tax rate from 22.1% to 15% beginning in 2012.[59]

U.S. Markets

Oil sand producers continue to look to the United States for the majority of their exports. Seventy-five percent of Canadian nonconventional oil exported to the United States is delivered to the Petroleum Administration for Defense District (PADD)[60] II in the Midwest. This region is well positioned to receive larger volumes of nonconventional oil from Canada because of its refinery capabilities. Several U.S.-based refinery expansions have been announced that would come online between 2007-2015. If Canada were to reach its optimistic forecasted oil sands output level of 5 mbd in 2030, and maintained its export level to the United States at around 90%, it would be exporting about 4.5 mbd to the United States. This would mean that imports from Canada would reach nearly 30% of all U.S. crude oil imports. U.S. refinery capacity is forecast to increase from 16.9 mbd in 2004 to nearly 19.3 mbd in 2030,[61] a 2.4 mbd increase — significant but perhaps not enough to accommodate larger volumes of oil from Canada, even if refinery expansions would have the technology to process heavier oil blends. Canada is pursuing additional refinery capacity for its heavier oil.

Pipelines. Oil sands are currently moved by two major pipelines (the Athabasca and the Corridor, not shown in Figure 7) as diluted bitumen to processing facilities in Edmonton. After reaching refineries in Edmonton, the synthetic crude or bitumen is moved by one of several pipelines to the United States (see Figure 7). The Athabasca pipeline has capacity of 570,000 barrels per day (b/d) while the Corridor has capacity of less than 200,000 b/d. Current pipeline capacity has nearly reached its limit. However, there are plans to increase Corridor's capacity to 610,000 b/d by 2010.

A number of new pipeline projects have been proposed or initiated that would increase the flow of oil from Canada to the United State's PADDs II, III, and V. Most of the new projects are scheduled to come online between 2008 and 2012. In addition, a couple of U.S. pipelines reversed their flow of crude oil (from south to north) to now carry Canadian heavy crude, originating from oil sands, to Cushing Oklahoma and Southeast Texas. Pipeline capacity could be a constraint to growth in the near term but the NEB predicts some excess pipeline capacity by 2009. An estimated $31.7 billion has been invested in pipeline projects for oil sands in western Canada.[62]

Environmental and Social Issues

The Federal Government of Canada classified the oil sands industry as a large industrial air pollution emitter (i.e., emitting over 8,000 tons CO_2/year) and expects it to produce half of Canada's growth in greenhouse gas (GHG) emissions[63] (about 8% total GHG emissions) by 2010. The oil sands industry has reduced its "emission emissions intensity" by 29% between 1995-2004 while production was rising. CO_2 have declined from 0.14 tons/bbl to about 0.08 tons/bbl or about 88 megatons since 1990.[64] Alberta's GHG goals of 238 megatons of CO_2 in 2010, and 218 megatons CO_2 in 2020 are not expected to be met.[65] Reducing air emissions is one of the most serious challenges facing the oil sands industry. However, according to the Pembina Institute, a sustainable energy advocate, greenhouse gas emissions intensity (CO_2/barrel) from oil sands is three times as high as that from conventional oil production.[66] The industry believes if it can reduce energy use it can reduce its emissions. As emissions per barrel of oil from oil sands decline overall, the Canadian government projects that total GHG emissions will continue to rise through 2020, attributing much of the increase to increased oil sands production.[67]

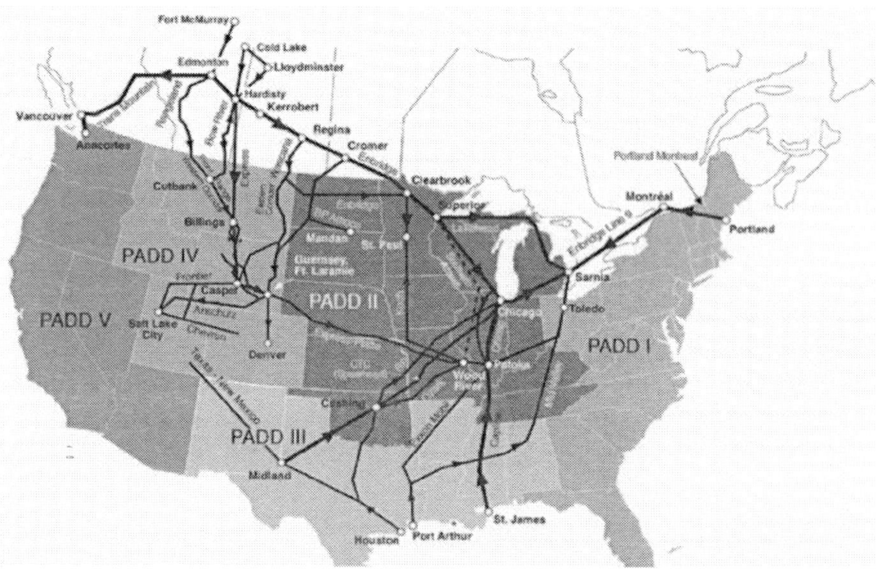

Source: *Canada's Oil Sands, Opportunities and Challenges to 2015: An Update*, June 2006.

Figure 7. Major Canadian and U.S. (Lower 48) Crude Oil Pipelines and Markets

Water supply and waste water disposal are among the most serious concerns because of heavy use of water to extract bitumen from the sands. For an oil sands mining operation, about 2-3 barrels of water are used from the Athabasca river for each barrel of bitumen produced; but when recycled produced water is included, 0.5 barrels of "make-up" water is required, according to the Alberta Department of Energy. Oil sands projects currently divert 150 million cubic meters of water annually from the Athabasca River but are approved to use up to 350 million cubic meters.[68] Concerns, however, arise over the inadequate flow of the river to maintain a healthy ecosystem and meet future needs of the oil sands industry. Additionally, mining operations impact freshwater aquifers by drawing down water to prevent pit flooding.

The freshwater used for in-situ operations is needed to generate steam, separate bitumen from the sand, hydrotransport the bitumen slurry, and upgrade the bitumen to a light crude. For SAGD operations, 90-95% of all the water used is recycled. Since some water is lost in the treatment process, additional freshwater is needed. To minimize the use of new freshwater supplies, SAGD operators use saline water from deeper underground aquifers. The use of saline water, however, generates huge volumes of solid waste which has posed serious disposal problems.

Wastewater tailings (a bitumen, sand, silt, and fine clay particles slurry) also known as "fluid fine tailings" are disposed in large ponds until the residue is used to fill mined-out pits. Seepage from the disposal ponds can result from erosion, breaching, and foundation creep.[69] The principal environmental threat is the migration of tails to a groundwater system and leaks that might contaminate the soil and surface water.[70] The tailings are expected to reach 1 billion cubic meters by 2020. Impounding the tailings will continue to be an issue even after efforts are made to use alternative extraction technology that minimizes the amount of tails. Tailings management criteria were established by the Alberta Energy and Utilities Board/Canadian Environmental Assessment Agency in June 2005. Ongoing extensive research by the Canadian Oil Sands Network for Research and Development (CONRAD) is focused on the consolidation of wastewater tailings, detoxifying tailings water ponds, and reprocessing tailings. Some R&D progress is being made in the areas of the cleanup and reclamation of tailings using bioremediation and electrocoagulation.[71]

The National Research Council of Canada (NRC) is conducting research to treat wastewater tailings and recover their byproduct residual bitumen, heavy metals, and amorphous solids (fertilizers). A pilot project is underway to clean and sort tailings, and recover metals such as aluminum and titanium.[72]

Surface disturbance is another major issue. The oil sands industry practice leaves land in its disturbed state and left to revegetate naturally. Operators, however, are responsible over the long term to restore the land to its previous potential.[73] Under an Alberta Energy Utility Board directive (AEUB), Alberta's Upstream Oil and Gas Reclamation and Remediation Program has expanded industry liability for reclaiming sites. The directive requires a "site-specific liability assessment" that would estimate the costs to abandon or reclaim a site.[74]

The government of Alberta's Department of the Environment established a "Regional Sustainable Development Strategy" whose purpose is, among other things, to "ensure" implementation of management strategies that address regional cumulative environmental impacts.[75] The oil sands industry is regulated under the Environmental Protection and Enhancement Act, Water Act, and Public Lands Act. Oil sands development proposals are reviewed by AEUB, Alberta Environment, and the Alberta Sustainable Resource Development at the provincial level. Review at the federal level may also occur.

ISSUES FOR CONGRESS

The Energy Policy Act of 2005 (P.L. 109-58) describes U.S. oil sands (along with oil shale and other unconventional fuels) as a strategically important domestic resource "that should be developed to reduce the growing dependence of the United States on politically and economically unstable sources of foreign oil imports."[76] The provision also requires that a leasing program for oil sands R&D be established. Given U.S. oil sands' strategic importance, but limited commercial success as discussed above, what level of federal investment is appropriate to reach U.S. energy policy goals? While an estimated 11 billion barrels of U.S. oil sands may be significant if it were economic, it represents a small share of the potentially recoverable resource base of unconventional fuels (e.g., 800 billion barrels of potentially recoverable oil from oil shale and another 20 billion barrels of recoverable heavy oil). Where is the best return on the R&D dollar invested for increased domestic energy supply and what are the long-term prospects for commercial application of unconventional fuels technology? Another important consideration to look at is where the oil industry is investing its capital and R&D for oil sands projects.

In light of the environmental and social problems associated with oil sands development, e.g., water requirements, toxic tailings, carbon dioxide emissions, and skilled labor shortages, and given the fact that Canada has 175 billion barrels of reserves and a total of over 300 billion barrels of potentially recoverable oil sands (an attractive investment under current conditions demonstrated by the billions of dollars already committed to Canadian development), the smaller U.S. oil sands base may not be a very attractive investment in the near-term.

U.S. refinery and pipeline expansions are needed to accommodate Canadian oil sands developments. Those expansions will have environmental impacts, but the new infrastructure could strengthen the flow of oil from Canadian oil sands. This expanded capacity will likely lead to even greater investment in Canada.

Whether U.S. oil sands are developed, Congress will continue to be faced with regulatory matters. Oil imports from oil sands are likely to increase from Canada and the permitting of new or expanded oil refineries will continue to be an issue because of the need to balance concerns over the environment on one hand and energy security on the other.

PROSPECTS FOR THE FUTURE

Because capital requirements for oil sands development has been enormous and risky, government involvement was seen as being essential in Canada, particularly during sustained periods of low oil prices. This private sector/government partnership in R&D, equity ownership, and public policy initiatives over the last 100 years has opened the way for the current expansion of the oil sands industry in Alberta.

Ongoing R&D efforts by the public and private sectors, sustained high oil prices, and favorable tax and royalty treatment are likely to continue to attract the increasing capital expenditures needed for growth in Canada's oil sands industry. Planned pipeline and refinery expansions and new upgrading capacity are underway to accommodate the increased volumes of oil sands production in Canada. U.S. markets will continue to be a major growth area for oil production from Canadian oil sands. Currently, about 5% of the total oil refined in the United States is from Canada's oil sands.

Even though prospects for Canadian oil sands appear favorable, factors such as water availability, waste water disposal, air emissions, high natural gas costs, insufficient skilled labor, and infrastructure demands may slow the pace of expansion.

Prospects for commercial development of U.S. oil sands are uncertain at best because of the huge capital investment required and the relatively small and fragmented resource base. The Task Force on Strategic Unconventional Fuels reported that oil sands comprise only about 0.6% of U.S. solid and liquid fuel resources, while oil shale accounts for nearly 25% of the total resource base.[77]

Appendix A

Table A1. Estimated World Oil Resources (in billions of barrels)

Region and Country	Proved Reserves	Reserve Growth	Undiscovered	Total
OECD				
United States	22.4	76.0	83.0	180.4
Canada[78]	178.8	12.5	32.6	223.8
Mexico	12.9	25.6	45.8	84.3
Japan United States	0.1	0.1	0.3	0.5
Australia/ New Zealand	1.5	2.7	5.9	10.1
OECD Europe	15.1	20.0	35.9	71.0
Non-OECD				
Russia	60.0	106.2	115.3	281.5
Other Non-OECD Europe/Eurasia	19.1	32.3	55.6	107.0
China	18.3	19.6	14.6	52.5
India	5.8	3.8	6.8	16.4
Other Non-OECD Asia	10.3	14.6	23.9	48.8
Middle East	743.4	252.5	269.2	1,265.1
Africa	102.6	73.5	124.7	300.8
Central and South America	103.4	90.8	125.3	319.5
Total	1,292.5	730.2	938.9	2,961.6
OPEC	901.7	395.6	400.5	1,697.8
Non-OPEC	390.9	334.6	538.4	1,263.9

Sources: Proved Reserves as of January 1, 2006: *Oil & Gas Journal*, vol. 103, no. 47 (December 19, 2005), p. 46-47. Reserve Growth Total and Undiscovered, 1995-2025; U.S. Geological Survey, *World Petroleum Assessment 2000*, website [http://pubs.usgs.gov/dds/dds-060/]. Estimates of Regional Reserve Growth: Energy Information Administration, *International Energy Outlook 2006*, DOE/EIA-0484(2006) (Washington, DC, June 2006), p. 29.

Note: Resources Include crude oil (including lease condensates) and natural gas plant liquids.

APPENDIX B

Table B1. Regional Distribution of Estimated Technically Recoverable Heavy Oil and Natural Bitumen (in billions of barrels)

Region	Heavy Oil		Natural Bitumen (oil sands)	
	Recovery Factor[a]	Technically Recoverable	Recovery Factor[a]	Technically Recoverable
North America	0.19	35.3	0.32	530.9
South America (Venezuela)	0.13	265.7	0.09	0.1
W. Hemisphere	0.13	301.0	0.32	531.0
Africa	0.18	7.2	0.10	43.0
Europe	0.15	4.9	0.14	0.2
Middle East	0.12	78.2	0.10	0.0
Asia	0.14	29.6	0.16	42.8
Russia	0.13	13.4	0.13	33.7[b]
E. Hemisphere	0.13	133.3	0.13	119.7
World		434.3		650.7

Source: U.S. Department of the Interior. U.S. Geological Survey Fact Sheet, FS 070-03 August 2003.

Note: Heavy oil and natural bitumen are resources in known accumulations.

a. Recovery factors were based on published estimates of technically recoverable and in-place[79] oil or bitumen by accumulation. Where unavailable, recovery factors of 10% and 5% of heavy oil or bitumen in-place were assumed for sandstone and carbonate accumulations, respectively.

b. In addition, 212.4 billion barrels of natural bitumen in-place is located in Russia but is either in small deposits or in remote areas in eastern Siberia.

ACRONYMS AND ABBREVIATIONS

AEUB	Alberta Energy and Utility Board
API	American Petroleum Institute
ARC	Alberta Research Council
ARCO	Atlantic Richfield Company
CCA	Capital Cost Allowance
CONRAD	Canadian Oil Sands Network for Research and Development

COS	Canadian Oil Sands
CSS	Cyclic Steam Stimulator
EIA	Energy Information Administration
GCOS	Great Canadian Oil Sands Company
GHG	greenhouse gases
IEA	International Energy Agency
mbd	million barrels per day
NEB	National Energy Board
OPEC	Organization of Petroleum Exporting Countries
PADD	Petroleum Administration for Defense District
R&D	research and development
ROI	return on investment
SAGD	steam-assisted gravity drainage
SCO	synthetic crude oil
SIRCA	Scientific and Industrial Research Council of Alberta
USGS	United States Geological Survey
VAPEX	Vapor Extraction Process

End Notes

[1] DOE, EIA, *International Energy Outlook, 2006*, p. 29.

[2] U.S. Department of Energy, EIA, *Annual Energy Outlook*, 2006.

[3] Oil Sands Fever, *The Environmental Implications of Canada's Oil Sand Rush*, by Dan Woynillowicz, et. al, The Pembina Institute, November 2005.

[4] Reserves are defined by the EIA as estimated quantities that geological and engineering data demonstrate with reasonable certainty to be recoverable in future years from known reservoirs under existing economic and operating conditions. Resources are defined typically as undiscovered hydrocarbons estimated on the basis of geologic knowledge and theory to exist outside of known accumulations. Technically recoverable resources are those resources producible with current technology without consideration of economic viability.

[5] In-situ mining extracts minerals from an orebody that is left in place.

[6] *Canada's Oil Sands: Opportunities and Challenges to 2015, An Energy Market Assessment*, National Energy Board, Canada, May 2004, p. 5.

[7] API represents the American Petroleum Institute method for specifying the density of crude petroleum. Also called API gravity.

[8] Diluents are usually any lighter hydrocarbon; e.g., pentane is added to heavy crude or bitumen in order to facilitate pipeline transport.

[9] *Canada's Oil Sands,* May 2004, p. 10.

[10] *Canada's Oil Sands*, May 2004, p. 4

[11] DOE, EIA, *International Energy Outlook, 2006*, p. 29

[12] *Canada's Oil Sands*, NEB, June 2006.

[13] *World Energy Investment Outlook, 2003 Insights*, International Energy Agency (IEA), 2003.

[14] Canadian Oil Sands, May 2004, p. 25.

[15] The U.S.-Canadian dollar exchange rate fluctuates daily. As of early October 2007 the exchange rate is U.S.$1 = C$0.9969. In December 2006 the exchange rate was U.S.$1 = C$1.15.

[16] *Oil Industry Update*, Alberta Economic Development, Spring 2005.

[17] Oil shale is a compact rock (shale) containing organic matter capable of yielding oil.

[18] U.S. Tar-Sand Oil Recovery Projects — 1984, L.C. Marchant, Western Research Institute, Laramie, WY, p. 625.

[19] Hydrocarbon-wetted oil sand deposits require different technology for bitumen extraction than that used for Alberta's water-wetted deposits. Oil sands are characterized as having a wet interface between the sand grain and the oil coating; this allows for the separation of oil from the grain. U.S. oil sands do not have a wet interface making the separation difficult.

[20] Phone communication with B. Tripp, Geologist, Utah Geological Survey, May 2004.

[21] Phone communication with Richard Meyers, Department of the Interior specialist in oil sands, September 2004.

[22] Presentation by Earth Energy Resources, Inc., at the Western U.S. Oil Sands Conference, University of Utah, September 21, 2006.

[23] *Development of America's Strategic Unconventional Fuels Resources, Initial Report to the President and the Congress of the United States*, Task Force on Strategic Unconventional Fuels, September 2006.

[24] The measured pace is based on sufficient private investment capital as a result of government policies but little direct government investment. An accelerated pace would imply a global oil supply shortage and rely more on significant government investment.

[25] *Development of America's Strategic Unconventional Fuels Resources, Initial Report to the President and the Congress of the United States*, Reference no. 17.

[26] The ARC was established in 1921, housed at the University of Alberta in Edmonton, and funded by the provincial government of Alberta. Its mandate was to document Alberta's mineral and natural resources. Today, the ARC is a wholly-owned subsidiary of the Alberta Science and Research Authority (ASRA) within Alberta's Ministry of Innovation and Science. The ARC has an annual budget of $85 million.

[27] The Influence of Interfacial Tension in the Hot-Water Process for Recovering Bitumen From the Athabasca Oil Sands, by L.L. Schramm, E.N. Stasiuk, and D. Turner, presented at the Canadian International Petroleum Conference, paper 2001-136, June 2001.

[28] Syncrude Canada Ltd. when first organized as a consortium of major oil companies comprised: Imperial Oil (an affiliate of Exxon), Atlantic Richfield (ARCO), Royalite Oil (later combined with Gulf Canada), and Cities Services R&D (See *The Syncrude Story*, p. 5). Its ownership has changed over the years as indicated in the text. Its current ownership structure is as follows: Canadian Oil Sands Ltd. (31.74%), Imperial Oil (25%), Petro-Canada Oil and Gas (12%), Conoco Phillips Oil Sands Partnership II (9.03%), Nexen Inc. (7.23%), Murphy Oil Co. Ltd. (5%), Mocal Energy Ltd. (5%) and the Canadian Oil Sands Limited Partnership (5%).

[29] GCOS, Ltd., was later renamed Suncor.

[30] *A Billion Barrels for Canada, The Syncrude Story*, pp. 44-45.

[31] The Alberta Energy Company (AEC) was created by the government of Alberta in 1975. Fifty percent was publicly owned. The government phased out is equity interest and in 1993 sold its remaining interest. The AEC and PanCanadian Energy Corporation merged in 2002 and became EnCana. EnCana sold its interest in Syncrude in 2003. For more details see Alexander's Oil and Gas Connection, "Company News North America," January 15, 2004.

[32] *The Syncrude Story*, pp. 72-73.

[33] Ibid, p. 98-99.

[34] Ibid, p.104

[35] Ibid, p. 122.

[36] Ibid, p. 136.

[37] The Alberta Energy Research Institute: Strategic Research Plan, 2003.

[38] ARC, Guide to the ARC, 2001-02. The ARC's more recent focus on developing in-situ technologies is beginning to shift back to surface mining R&D. They believe that their role is to help many of the newcomers to the industry develop "best practices" technology. The

ARC sees itself as an ongoing player in the R&D business because of the huge challenges related to environmental quality, cost reductions, and the need for new upgrading technologies and refinery expansions.

[39] *Oil Sands Technology Roadmap*, p. 20.

[40] *Oil Sands Industry Update*, AED, June 2006, p. 7.

[41] *Oil Sands Technology Roadmap: Unlocking the Potential*, Alberta Chamber of Resources, January 2004 p. 23.

[42] According to the National Energy Board Report, one thousand cubic feet of natural gas is required per barrel of bitumin for SAGD operations. *Canada's Oil Sands,* May 2004.

[43] *Canada's Oil Sands,* June 2006 p. 4.

[44] *Canada's Oil Sands, Opportunities and Challenges to 2015, An Energy Market Assessment*, May 2004, National Energy Board, Canada, p. 108.

[45] *Canada's Oil Sands*, June 2006.

[46] *Overview of Canada's Oil Sands*, TD Securities, January 2004, p. 19.

[47] NEB, June 2006, pp. 20-21.

[48] COS, 2004, p. 9.

[49] COS, 2006, p. 5.

[50] COS, 2004, p. 18.

[51] The generic oil sands royalty regime consists of three parts: the lease sale, a minimum 1% pre-payout gross revenue royalty, and a 25% post-payout net revenue royalty. The payout period is the time it takes a firm to recover all allowable capital costs including a rater of return.

[52] *Oil and Gas Fiscal Regime, Alberta Resource Development of Western Canadian Provinces and Territories*, p. 39, 1999.

[53] *Oil Sands, Benefits to Alberta and Canada, Today and Tomorrow, Through a Fair, Stable and Competitive Fiscal Regime*, Canadian Association of Petroleum Producers, May 2007, Appendix B.

[54] *Our Fair Share*, Report of the Alberta Royalty Review Panel, September 18, 2007, p. 7.

[55] Ibid., p. 27.

[56] *Overview of Canada's Oil Sands*, T.D. Securities, January 2004, p. 7.

[57] *The New Royalty Framework*, October 25, 2007.

[58] *Oil and Gas Taxation in Canada*, January 2000, PriceWaterhouseCoopers.

[59] Canadian Department of Finance, Economic Statement, October 30, 2007.

[60] There are 5 PADD's in the United States. PADDs were created during World War II as a way to organize the distribution of fuel in the United States.

[61] DOE/EIA, *Annual Energy Outlook 2006 with Projections to 2030*, February 2006.

[62] "Oil Sands Producers Facing Pipeline Capacity Constraints," *The Energy Daily,* August 7, 2007.

[63] Greenhouse gas emissions include carbon dioxide, methane, nitrous oxide, hydrofluorocarbons, perfluorocarbons, and sulfur hexafluoride.

[64] COS, 2004, p. 62.

[65] Ibid., p. 63.

[66] *Oil Sands Fever*, by Dan Woynillowicz, et al., The Pembina Institute, November 2005.

[67] COS, June 2006, p. 39.

[68] *Oil Sands Fever*, op. cit.

[69] *Canada's Oil Sands* (water conservation initiatives), pp. 66-68.

[70] *Canada's Oil Sands*, p. 68.

[71] Ibid., p. 69.

[72] For more on byproducts, see *Canada's Oil Sands*, p. 70.

[73] Ibid., p. 71.

[74] Ibid.

[75] *Oil Sands Industry Update,* AED, June 2006, p. 29.

[76] Section 369 of Energy Policy Act of 2005.

[77] *Development of America's Strategic Unconventional Fuels Resources,* September 2006, p. 5.

[78] Oil sands account for 174 billion barrels of Canada's total 179 billion barrel oil reserves. Further, the Alberta Energy and Utilities Board estimates that Alberta's oil sands contain 315 billion barrels of ultimately recoverable oil. *Canada's Oil Sands: Opportunities and Challenges to 2015: An Update, June 2006*, National Energy Board.

[79] In-place oil is a continuous ore body that has maintained its original characteristics.

In: Canadian Oil Sands, Greenhouse Gases ... ISBN: 978-1-62257-402-5
Editors: T. J. Carlson and S. Harper

Chapter 3

KEYSTONE XL PIPELINE PROJECT: KEY ISSUES*

Paul W. Parfomak, Neelesh Nerurkar, Linda Luther and Adam Vann

SUMMARY

In 2008, Canadian pipeline company TransCanada filed an application with the U.S. Department of State to build the Keystone XL pipeline, which would transport crude oil from the oil sands region of Alberta, Canada, to refineries on the U.S. Gulf Coast. Keystone XL would ultimately have the capacity to transport 830,000 barrels per day, delivering crude oil to the market hub at Cushing, OK, and further to points in Texas. TransCanada plans to build a pipeline spur so that oil from the Bakken formation in Montana and North Dakota can also be carried on Keystone XL.

As a facility connecting the United States with a foreign country, the pipeline requires a Presidential Permit from the State Department. In evaluating such a permit application, the department must determine whether it is in the "national interest," considering the project's potential effects on the environment, economy, energy security, foreign policy, and other factors. Environmental impacts are considered pursuant to the National Environmental Policy Act, and documented by the State

* This is an edited, reformatted and augmented version of Congressional Research Service, Publication No. R41668, dated May 9, 2012.

Department in an Environmental Impact Statement (EIS). The final EIS was released for the Keystone XL pipeline permit application in August 2011, after which a 90- day public review period began to make the national interest determination. During that time the State Department determined that more information was needed to consider an alternative pipeline route avoiding the environmentally sensitive Sand Hills region of Nebraska, an extensive sand dune formation with highly porous soil and a shallow depth to groundwater recharging the Ogallala aquifer.

The Temporary Payroll Tax Cut Continuation Act of 2011 (P.L. 112-78) required the Secretary of State to approve or deny the project within 60 days. On January 18, 2012, the State Department, with the President's consent, denied the Keystone XL permit, citing insufficient time under this deadline to properly assess the reconfigured project. Subsequently, TransCanada announced that it would proceed with development of the pipeline segment connecting Cushing, OK, to the Gulf Coast as a stand-alone project not requiring a Presidential Permit—a decision supported by the Obama administration. In April 2012, TransCanada submitted to Nebraska proposed pipeline routes avoiding the Sand Hills. Subsequently, on May 4, 2012, TransCanada submitted a new application for a Presidential Permit that includes proposed new routes through Nebraska. With the new permit application, the NEPA compliance process begins anew, although it may draw from relevant existing analysis and documentation prepared for the August 2011 final EIS.

In the wake of the State Department's denial of the Presidential Permit, Congress has debated legislative options addressing the Keystone XL pipeline. The Surface Transportation Extension Act of 2012, Part II (H.R. 4348) and the North American Energy Access Act (H.R. 3548) would transfer the permitting authority for the Keystone XL pipeline project to the Federal Energy Regulatory Commission, requiring FERC to issue a permit within 30 days of enactment. The Keystone For a Secure Tomorrow Act (H.R. 3811), the Grow America Act of 2012 (S. 2199), S. 2041 (a bill to approve the Keystone XL pipeline), the EXPAND Act (H.R. 4301), and the Energizing America through Employment Act (H.R. 4000) would immediately approve the original permit application filed by TransCanada. All seven bills include provisions allowing for later alteration of the pipeline route in Nebraska. S. 2100 and H.R. 4211 would suspend sales of petroleum products from the Strategic Petroleum Reserve until issuance of a Presidential Permit for the Keystone XL project. Changing or eliminating the State Department's role in issuing cross-border infrastructure permits may raise questions about the President's executive authority, however. H.R. 3900 would seek to ensure that crude oil transported by the Keystone XL pipeline, or resulting refined petroleum products, would be sold only into U.S. markets, but this bill could raise issues related to international trade agreements.

INTRODUCTION[1]

In September 2008, TransCanada (a Canadian company) applied to the U.S. Department of State for a permit to cross the U.S.-Canada international border with the Keystone XL pipeline project. If constructed, the pipeline would carry crude oil produced from the oil sands region of Alberta, Canada, to U.S. Gulf Coast refineries. Because the pipeline would connect the United States with a foreign country, it requires a Presidential Permit issued by the State Department. Issuance of a Presidential Permit requires a finding that the project would serve the "national interest."

In the course of gathering information necessary to make its national interest determination, the State Department identified various concerns raised by the public. On November 10, 2011, the State Department announced its decision to seek additional information about alternative pipeline routes before it could move forward with a national interest determination.[2] More specifically, concerns regarding potential environmental impacts of constructing and operating the pipeline along the proposed route through the Sand Hills region of Nebraska led the State Department to decide that an assessment of potential alternative routes that would avoid that area was necessary. Subsequently, on November 14, 2011, TransCanada announced an agreement with the Nebraska Department of Environmental Quality to identify a pipeline route that would avoid the Sand Hills. The State Department estimated at the time that the preparation of supplemental environmental analysis necessary for a new route alternative could be complete in early 2013.

After the State Department's announcement of a delay in the permit review, Congress acted to expedite a permit decision on the Keystone XL project. The Temporary Payroll Tax Cut Continuation Act of 2011 (P.L. 112-78), enacted on December 23, 2011, included provisions requiring the Secretary of State to issue a permit for the project within 60 days, unless the President publicly determined the project not to be in the national interest. The act allowed for future changes to the Nebraska route if approved by the governor of Nebraska. On January 18, 2012, the State Department, with the President's consent, denied the Keystone XL permit, citing insufficient time under the 60-day deadline to obtain all the necessary information to assess the reconfigured project.[3]

On February 27, 2012, TransCanada announced that it would proceed with development of the Keystone XL segment connecting Cushing, OK, to the Gulf Coast as a stand-alone project not requiring a Presidential Permit.[4] The company also informed the State Department that it intended to file a new

Presidential Permit application "in the near future" for the segment of the Keystone XL project from the Canadian border to Steele City, NE, with a future supplement to the application specifying an alternative route in Nebraska when that route is selected.[5] On May 4, 2012, the State Department received a new application from TransCanada for a proposed pipeline that would run from the Canadian border to connect to an existing pipeline in Steele City, NE.[6] The new application includes proposed new routes through Nebraska. The Obama administration has stated its support for TransCanada's plan to proceed with the southernmost segment of the Keystone XL pipeline, while reserving judgment on the reconfigured northern segment until completion of a new Presidential Permit review.[7] Members of Congress have long expressed support for the proposed pipeline's potential energy security and economic benefits, while others have expressed reservations about its potential environmental impacts.[8] Though Congress, to date, has had no direct role in permitting the pipeline's construction, it may have an oversight role stemming from federal environmental statutes that govern the pipeline's application review process. Congress also may seek to influence the State Department permitting process, or may seek to assert direct congressional authority over permit approval, through new legislation. A number of legislative proposals, like P.L. 112-78, would have imposed deadlines on a national interest determination for the Keystone XL project. The North American-Made Energy Security Act (H.R. 1938), would have directed the President to issue a final order granting or denying the Presidential Permit for the Keystone XL pipeline by November 1, 2011. The Jobs Through Growth Act (H.R. 3400), would have required the President to issue a final order granting or denying the Presidential Permit for the Keystone XL pipeline within 30 days of enactment. The North American Energy Security Act (S. 1932), which was introduced on November 30, 2011, would have required the Secretary of State to issue a permit for the project within 60 days of enactment, unless the President publicly determined the project to be not in the national interest. The North American Energy Security Act (H.R. 3537), introduced on December 1, 2011, and the Middle Class Tax Relief and Job Creation Act of 2011 (H.R. 3630), introduced on December 9, 2011, contain similar provisions for issuing a Presidential Permit within 60 days of enactment. All of these proposed provisions were mooted by the State Department's initial denial of the permit. Provisions in the Surface Transportation Extension Act of 2012, Part II (H.R. 4348), which passed in the House on April 18, 2012, and in the North American Energy Access Act (H.R. 3548), introduced on December 2, 2011, would transfer the permitting authority over the Keystone XL pipeline project from the State Department to the Federal Energy Regulatory Commission (FERC), requiring the commission to issue a permit for the project within 30 days of enactment. The Keystone For a Secure Tomorrow Act (H.R. 3811), introduced on January 24, 2012, would immediately approve the original permit application filed by TransCanada in 2008. The Grow America Act of 2012 (S. 2199), introduced on March 15, 2012, the EXPAND Act (H.R. 4301), introduced on March 29, 2012, S. 2041 (a bill to approve the Keystone XL pipeline), introduced on January 30, 2012, and the Energizing America through Employment Act (H.R. 4000), introduced on February 9, 2012, would similarly approve the original permit upon passage. All seven bills include provisions allowing for later alteration of the pipeline route in Nebraska.

H.R. 3900, introduced on February 3, 2012, would seek to ensure that any crude oil transported by the Keystone XL pipeline, or resulting refined petroleum products, would be sold only into U.S. markets—not exported overseas. S. 2100, introduced on February 13, 2012, and H.R. 4211, introduced on March 19, 2012, would suspend sales of petroleum products from the Strategic Petroleum Reserve until issuance of a Presidential Permit for the Keystone XL project application filed in 2008.

This report describes the Keystone XL pipeline proposal and the process required for federal approval. It summarizes key arguments for and against the pipeline put forth by the pipeline's developers, federal agencies, environmental groups, and other stakeholders. Finally, the report reviews the constitutional basis for the State Department's authority to issue a Presidential Permit, and opponents' possible challenges to this authority.

Pipeline Description

The U.S. portion of the Keystone XL pipeline project, as originally proposed, would pass through Montana, South Dakota, Nebraska, Oklahoma, and Texas (Figure 1). This route would consist of approximately 1,380 miles of 36-inch-diameter pipe and have the capacity to transport 830,000 barrels per day (bpd) of crude oil to the United States, delivering up to roughly 200,000 bpd to an existing oil terminal in Oklahoma with the remainder sent further to delivery points in Texas.[9] On November 14, 2011, TransCanada announced an agreement with the state of Nebraska to make as yet undetermined changes to the pipeline route in Nebraska to avoid environmentally sensitive areas.[10] These route changes are expected to increase the pipeline mileage through the state, although the company expects to maintain the original route through the other states, including its planned delivery points, so the pipeline's overall capacity should not be affected.

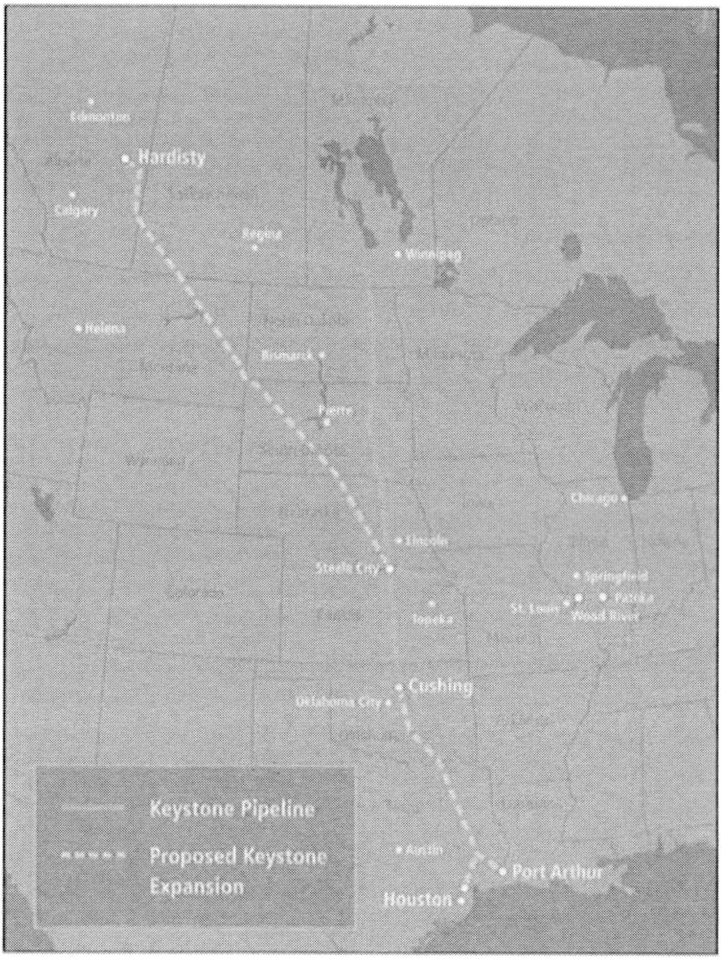

Source: U.S. Department of State, http://keystonepipeline-xl.state.gov/clientsite/keystonexl.nsf/map.jpg? OpenFileResource.

Note: Figure 1 shows the developer's originally proposed "preferred alternative" for the Keystone XL pipeline route according to Presidential Permit application documents. However, the route through Nebraska will change. For discussion of alternative routes, see the State Department EIS discussed below and Figure 2.

Figure 1. TransCanada Keystone Pipeline and Original Keystone XL Proposed Route.

As of February 2011, the Keystone XL project along its original route was estimated to cost more than $7.0 billion, with the U.S. portion accounting for at least $5.4 billion of that total.[11] That is higher than the cost estimate when the initial permit application was filed, reportedly due to currency swings, changing regulatory requirements, and permitting delays.[12] A new route would presumably be longer and cost more. The Keystone XL pipeline would be an extension of TransCanada's existing Keystone pipeline, which links the Alberta oil sands to refineries in Illinois and Oklahoma (Figure 1). The Keystone pipeline received State Department approval on March 17, 2008, and began commercial operation in June 2010.

Keystone XL Extension to Bakken Oil Production

The U.S. portion of the Bakken formation is an unconventional oil resource that underlies parts of North Dakota and Montana.[13] By the end of 2010, U.S. Bakken production was 350,000 bpd.[14] Output has climbed further since then. The oil is transported to refineries by rail and truck, rather than by pipeline, which would be more economic. In part, this is because infrastructure has not kept up with rapid production growth in the Bakken region in recent years. Output is expected to increase significantly in the future, increasing the need for pipeline transportation capacity.[15]

TransCanada has signed contracts with Bakken oil producers to carry 65,000 bpd from the region via the Keystone XL pipeline. While not the full 100,000 bpd of capacity TransCanada had offered to oil producers, this was enough to justify adding the Bakken Marketlink Project, a pipeline running from Baker, MT, to the Keystone XL pipeline, which can then carry crude to the oil hub at Cushing, OK and on to the Gulf Coast.[16] The Bakken Marketlink would have a 100,000 bpd capacity and is estimated to cost $140 million.[17]

These new Bakken contracts also improve the economics for Keystone XL, raising the amount of oil slated to flow through the pipeline.[18] Lower transportation costs and access to new markets may support investment in the Bakken. Furthermore, TransCanada is not the only company adding pipeline capacity in the region. Notably, Enbridge, another Canadian pipeline company, has proposed the Bakken Pipeline Project, which will add 120,000 bpd of transport capacity to move Bakken oil to Midwest markets.[19] According to Enbridge, sufficient pipeline capacity has been slow to emerge in the region because "they're smaller players in the Bakken. They are not able to make the 20-year commitments and it's been a lot of work to get them to commit to the level that [is] required to underwrite a major project out of the Bakken."[20] Rail transport capacity is also expanding.[21]

PRESIDENTIAL PERMIT APPLICATION REQUIREMENTS

Ordinarily, federal agencies have no authority to site oil pipelines, even interstate pipelines.[22] The primary siting authority for oil pipelines generally would be established under applicable state law (which may vary considerably from state to state). However, the construction, connection, operation, and maintenance of a pipeline that connects the United States with a foreign country requires executive permission conveyed through a Presidential Permit. Since the Keystone and proposed Keystone XL pipelines are designed for the importation of oil from Canada, their facilities require a Presidential Permit.

Executive Order 13337 delegates to the Secretary of State the President's authority to receive applications for Presidential Permits.[23] Issuance of a Presidential Permit is dependant upon a finding that the project would serve the "national interest." In the course of making that determination, the State Department is obligated to consider a host of issues related to the proposed project. The State Department will not necessarily evaluate the same factors for each project seeking a permit. However, the State Department identified the following as key factors it considered in making previous national interest determinations for pipeline permit applications:

- Environmental impacts of the proposed projects;
- Impacts of the proposed projects on the diversity of supply to meet U.S. crude oil demand and energy needs;
- The security of transport pathways for crude oil supplies to the United States through import facilities constructed at the border relative to other modes of transport;
- Stability of trading partners from whom the United States obtains crude oil;
- Relationship between the United States and various foreign suppliers of crude oil and the ability of the United States to work with those countries to meet overall environmental and energy security goals;
- Impact of proposed projects on broader foreign policy objectives, including a comprehensive strategy to address climate change;
- Economic benefits to the United States of constructing and operating proposed projects; and
- Relationships between proposed projects and goals to reduce reliance on fossil fuels and to increase use of alternative and renewable energy sources.[24]

In making its national interest determination, the State Department is required to consult with relevant federal and state agencies and to invite public comment in arriving at its determination. However, the State Department has broad discretion in determining what factors it will examine to inform its determination and, ultimately, whether a proposed project is in the national interest.

Documenting Environmental Impacts under NEPA

As identified on the list above, a proposed project's environmental impacts is one factor considered by the State Department in making its national interest determination. The State Department's identification and consideration of environmental impacts is documented within the context of preparing an Environmental Impact Statement (EIS), pursuant to the National Environmental Policy Act (NEPA, 42 U.S.C. §4321 et seq.).[25]

Broadly, NEPA requires federal agencies to consider the environmental impacts of their actions before proceeding with them and to inform the public of those potential impacts. To ensure that environmental impacts are considered, an EIS must be prepared for major federal actions "significantly" affecting the environment.[26] With respect to the 2008 Presidential Permit application submitted by TransCanada, the State Department concluded that issuance of a permit for the proposed construction, connection, operation, and maintenance of the Keystone XL Pipeline and its associated facilities at the United States border would constitute a major federal action that may have a significant impact upon the environment within the meaning of NEPA.[27] For this reason, the State Department prepared an EIS to address reasonably foreseeable impacts from the proposed action and alternatives. Similarly, an EIS will have to be prepared for pipeline project for which the May 4, 2012, permit application was filed.

Overview of the NEPA Process for the XL Pipeline Project

Among other requirements, an EIS must include a statement of the purpose and need for an action, a description of all reasonable alternatives to meet that purpose and need, a description of the environment to be affected by those alternatives, and an analysis of the direct and indirect effects of the alternatives, including cumulative impacts.[28] Accordingly, the State Department EIS must demonstrate the review and consider potential environmental impacts of the entire pipeline (including the construction, operation, and maintenance of the pipeline and its associated facilities), not just the facilities at the border crossing.

As the NEPA compliance process for TransCanada's permit application has proceeded, it is important to understand the distinction between what is required under NEPA itself and what may be required pursuant to other environmental requirements identified within the context of the NEPA process. NEPA itself requires federal agencies to identify the environmental impacts of an action before proceeding with them and to involve the public in that process when environmental impacts are significant. In that process of identifying a proposed project's environmental impacts, within the context of preparing the EIS, the lead agency should identify any compliance obligations (licenses, permits, or approvals) established under additional state, tribal and federal law applicable to the portion of the project constructed in the United States (see "State Siting and Additional Environmental Requirements," below).

EIS preparation is done in two stages, resulting in a draft and final EIS. NEPA regulations require the draft EIS to be circulated for public and agency comment, followed by a final EIS that incorporates those comments.[29] Preparing the EIS is the responsibility of a designated "lead agency," in this case, the State Department. In developing the EIS, the State Department must rely to some extent on information provided by TransCanada. For example, TransCanada's original permit application included an Environmental Report which was intended to provide the State Department with sufficient information to understand the scope of potential environmental impacts of the project.[30]

In preparing the draft EIS, the lead agency must request input from "cooperating agencies," which include any agency with jurisdiction by law or with special expertise regarding any environmental impact associated with the project.[31] Cooperating agencies for the Keystone XL project (for the pipeline's first Presidential Permit application) were the U.S. Environmental Protection Agency (EPA); the Department of Transportation's Pipeline and Hazardous Materials Safety Administration (PHMSA), Office of Pipeline Safety (OPS); the Department of the Interior's Bureau of Land Management, U.S. Fish and Wildlife Service, and National Park Service; the U.S. Army Corps of Engineers; the U.S. Department of Agriculture's Farm Service Agency, Natural Resources Conservation Service, and Rural Utilities Service; the Department of Energy's Western Area Power Administration; and state environmental agencies.

Table 1. NEPA Milestones for TransCanada's 2008 Presidential Permit Application; Administrative, Congressional, State, and Company Actions That Affected the NEPA Process

Date	Party	Description
Sept. 2008	TransCanada	An application for a Presidential Permit is filed with the State Department to build and operate the Keystone XL Project; a "Preliminary Environmental Report" for the project is also submitted.
Apr. 16, 2010	State Department	Draft EIS for the proposed Keystone XL Pipeline project is released for public comment.
July 16, 2010	EPA	The agency rates the draft EIS as "Inadequate," noting that potentially significant impacts were not evaluated, that the additional information and analysis was needed, and that the draft EIS would need to be formally revised and again made available for public review.
Oct. 21, 2010	State Department	Secretary Clinton states that the State Department was "inclined to" approve the project. Critics of the project, including some Members of Congress, stated that the Secretary's statement appeared to prejudge its permit approval for the pipeline proposal as a foregone conclusion.
Apr. 15, 2011	State Department	Supplemental draft EIS issued.
June 6, 2011	EPA	The agency rates the supplemental draft EIS as having "Insufficient Information" and the action as having "Environmental Objections." EPA recommends additional analysis on a range of issues.
Aug. 26, 2011	State Department	Final EIS issued.
Aug. - Oct. 2011	State Department	The 90-day public review period for National Interest Determination begins; State Department holds public meetings in the six states through which the proposed pipeline would pass and in Washington, D.C.
Oct. 2011	Congress	Fourteen Members of Congress request the State Department Office of Inspector General (IG) to investigate the department's handling of the EIS and National Interest Determination for the Keystone XL project.
Oct. 24, 2011	Governor of Nebraska	The Governor calls the Nebraska legislature into a special session to determine if siting legislation can be crafted and passed for pipeline routing in Nebraska.
Nov. 4, 2011	State Department	IG announces it is initiating a special review to determine to what extent the Department and all other parties involved complied with Federal laws and regulations relating to the Keystone XL pipeline permit process.

Table 1. (Continued)

Date	Party	Description
Nov. 10, 2011	State Department	The agency announces that additional information will be needed regarding alternative pipeline routes that would avoid the Nebraska Sand Hills before National Interest Determination can be made. Officials suggest that analysis needed to prepare the supplemental EIS, including additional public comment, could be completed as early as the first quarter of 2013.
Nov. 14, 2011	TransCanada	The company announces that it will work with the Nebraska Department of Environmental Quality (DEQ) to identify a potential pipeline route that would avoid the Nebraska Sand Hills.
Nov. 22, 2011	Governor of Nebraska	The Governor signs legislation passed during the special session directing the Nebraska DEQ to work collaboratively with the State Department to gather information necessary for a supplemental EIS.
Nov. 2011	Nebraska DEQ/State Department	The agencies begin to negotiate a Memorandum of Understanding (MOU) regarding their collaboration on the supplemental EIS. Nebraska DEQ hires a contractor to delineate the "Sand Hills" region that alternative routes must avoid.
Dec. 23, 2011	Congress	The Temporary Payroll Tax Cut Continuation Act of 2011 (P.L. 112-78) is enacted, including provisions requiring the Secretary of State to issue a permit for the project within 60 days, unless the President determines the project is not in the national interest.
Jan. 18, 2012	State Department	The agency announces, with the President's consent, that it will deny the Keystone XL permit. It states that its decision was predicated on the fact that the 60-day deadline under P.L. 112-78 did not provide sufficient time to obtain information necessary to assess the current project's national interest.
Feb. 3, 2012	State Department	Formal permit denial issued; State Department and Nebraska DEQ suspend work on MOU regarding a supplemental EIS.
Feb. 9, 2012	State Department	The IG releases findings reporting that the State Department "did not violate its role as an unbiased oversight agency," among other specific findings regarding department's Keystone XL permit review process.
Apr. 19, 2012	TransCanada	The company submits to the Nebraska DEQ an initial analysis of alternative Keystone XL pipeline routes that avoid the Sand Hills.
May 4, 2012	TransCanada	A new Presidential Permit application is submitted to State Department reflecting new information regarding alternative pipeline routes through Nebraska. The NEPA process for the new project begins, potentially drawing upon relevant documents from the 2011 final EIS.

Source: The Congressional Research Service, based on a review of events during, and affecting, the NEPA process conducted for the 2008 Presidential permit application for the Keystone XL pipeline project.

In addition to its role as a cooperating agency, EPA is also required to review and comment publicly on the EIS and rate both the adequacy of the EIS itself and the level of environmental impact of the proposed project.[32] Rating the EIS takes place after the draft is issued. The EIS could be rated either "Adequate," "Insufficient Information," or "Inadequate." EPA's rating of a project's environmental impacts may range from "Lack of Objections" to "Environmentally Unsatisfactory." In rating the impact of the action itself, EPA would specify one of the following: "Lack of Objections," "Environmental Concerns," "Environmental Objections," or "Environmentally Unsatisfactory." The federal agency would then be required to respond to EPA's rating, as appropriate.

EPA's role in rating draft EISs had a significant impact on the NEPA process for TransCanada's 2008 Presidential Permit application. Major milestones in that NEPA process are listed in Table 1 (for more detail on the milestones listed, see Appendix B).

As illustrated in Table 1, following the release of the Keystone XL project's final EIS, a public review period began during which the proposed project's national interest determination could be made. As part of this process for the initial Keystone XL project, the State Department held public meetings in each of the six states through which the proposed pipeline would pass and in Washington, DC.[33] The meetings were intended to give members of the public additional opportunity to voice their opinions on issues they thought should be taken into account in determining whether granting or denying the Presidential Permit would be in the national interest. These additional public meetings were not part of the NEPA process. Considering the strong public interest in the pipeline proposal (both opposed and in favor), the public hearings were part of the State Department's national interest determination. During the review period, the State Department received input from state, local, and tribal officials as well as members of the public.

On November 10, 2011, after the public review period, the State Department issued a statement regarding the public comments and its response to those comments.[34] The State Department stated that it received comments on a wide range of issues including the Keystone XL project's potential impact on jobs, pipeline safety, health concerns, the societal impact of the project, and oil extraction in Canada. Concern regarding the proposed pipeline route through the Sand Hills area of Nebraska was identified as one of the most common issues raised. Comments regarding that pipeline route were consistent with the environmental impacts identified in the final EIS with regard to the unique combination of characteristics of the Sand Hills region (e.g., a high concentration of wetlands of special concern, a sensitive ecosystem, and extensive areas of very shallow groundwater). Further, the Nebraska legislature was convening a special session to consider the legislation that would establish regulations applicable to pipeline siting within the state.

Facing the prospect of new state pipeline siting regulations applicable to the Sand Hills, together with the concern about the Keystone XL pipeline's specific "preferred" route, the State Department announced that it needed additional information about alternative pipeline routes avoiding the environmentally sensitive Sand Hills area in Nebraska before moving forward with its national interest determination.[35] Although the State Department did not decide that environmental issues led to a determination that the proposed project was not in the national interest, environmental issues identified in the final EIS, and further stressed in public comments, led to its decision to delay that determination until it gathered this information. In a concurrent press release, President Obama stated

> Because this permit decision could affect the health and safety of the American people as well as the environment, and because a number of concerns have been raised through a public process, we should take the time to ensure that all questions are properly addressed and all the potential impacts are properly understood.[36]

Subsequently, TransCanada announced that it would work with the State Department and the Nebraska Department of Environmental Quality (DEQ) to conduct an environmental assessment to define the best location for Keystone XL pipeline in Nebraska. Further, the company stated that it would "cooperate with these agencies and provide them with the information they need to complete a thorough review that addresses concerns regarding the Sandhills region."[37]

The Role of Environmental Impacts in a National Interest Determination

Generally, after a final EIS is issued for a pipeline project that requires a Presidential Permit, a final project decision would be reflected in a "Record of Decision and National Interest Determination," issued by the State Department.[38] That document, required under elements of both NEPA and E.O. 11424, formalizes the selection of a project alternative.

However, after the 90-day public review period, instead of issuing that final decision document, the State Department announced its decision to gather additional information regarding the alternative pipeline routes. This scenario illustrates the distinctly different, yet interrelated requirements applicable to NEPA process, that culminates with the publication of a final record of decision (ROD) and the national interest determination that is necessary under E.O. 13337 (as amended). Under NEPA, the State Department (or any other federal agency considering an action) must fully assess the environmental consequences of an action and potential project alternatives *before* making a final decision. NEPA does not prohibit a federal action that has adverse environment impacts; it requires only that a federal agency be fully *aware of* and *consider* those adverse impacts before selecting a final project alternative. That is, NEPA is intended to be part of the decision-making process, not dictate a particular outcome. The State Department's national interest determination decision, however, does dictate a particular outcome—issuance of a Presidential Permit. That is, issuance of a Presidential Permit is predicated on the Secretary of State finding that the proposed project would serve the national interest. While NEPA does not prohibit federal actions with adverse environmental impacts, a project's adverse environmental impacts (as well as other factors) may lead the State Department to determine that it is not in the national interest.

Although no new decision deadline was established, State Department officials suggested that it would be "reasonable to expect that this process including a public comment period on a supplement to the final EIS consistent with NEPA could be completed as early as the first quarter of 2013."[39] In a prior press interview, President Obama also appeared to suggest that, notwithstanding the delegation of Presidential Permit authority to the State Department, he would be personally involved in the final decision on the Keystone XL Pipeline permit application.[40]

Presidential Permit Denial

As noted earlier, on December 23, 2011, the Temporary Payroll Tax Cut Continuation Act of 2011 was enacted (P.L. 112-78). Under Section 501, "Permit for Keystone XL pipeline," the Secretary of State was required to grant the Presidential Permit for the Keystone XL pipeline project within 60 days, unless the President determined that the pipeline would not be in the national interest. On January 18, 2012, the State Department announced, with the President's concurrence, that the Presidential Permit for the proposed Keystone XL Pipeline would be denied at that time because it was determined not to serve the national interest. That recommendation "was predicated on the fact that the Department does not have sufficient time to obtain the information necessary to assess whether the project, in its current state, is in the national interest."[41] However, the department also stated that its decision did not preclude TransCanada from reapplying for a Presidential Permit in the future.

The State Department announcement reiterated the concerns regarding the proposed route through the Sand Hills area of Nebraska and the need for additional information regarding alternative routes around that region before it could make a national interest determination. As it had previously, the State Department estimated that, based on prior projects of similar length and scope and after consultations with the State of Nebraska and TransCanada, it could complete the necessary review to make a decision by the first quarter of 2013.

In a briefing following the State Department permit denial, it was noted that a new permit application would trigger a new environmental review process under NEPA.[42] It was further noted, however, that regulations under both NEPA and the State Department internal procedures allow for the use of relevant, existing NEPA documentation and analysis. The final EIS completed in August 2011, as well as the body of information gathered to prepare that document, could potentially be used to prepare required NEPA documentation for a new permit application. The degree to which the previous documentation would be useful would depend on the degree to which the previously proposed Keystone XL pipeline would be similar to a newly proposed pipeline project.

On February 27, 2012, in the wake of the Presidential Permit denial, TransCanada advised the State Department of its intent to file a Presidential Permit application in the near future and subsequently supplement that application with an alternative route in Nebraska, as soon as that route is selected by the state.[43] (The company began work on developing alternative routes that would avoid the Sand Hills shortly after the State Department's November 11th announcement that it would need more information about alternative routes as a condition of making its national interest determination.) TransCanada also advised the State Department that it would proceed with development of the pipeline segment connecting Cushing, OK, to the Gulf Coast as a standalone project (the Gulf Coast Pipeline Project) entirely within the United States. As a project that does not involve an international border crossing, a Presidential Permit will not be required for the Gulf Coast leg of the project. However, as a pipeline carrying a hazardous material, it will still be required to comply with a range of state and federal permit, approval, and consultation requirements (see "State Siting and Additional Environmental Requirements").

The New Permit Application Process

On April 19, 2012, the Nebraska DEQ received TransCanada's *Initial Report Identifying Alternative and Preferred Corridors for Nebraska Reroute* route.[44] Public meetings on the newly proposed routes were scheduled for May 9-17. On May 4, 2012, TransCanada submitted a new application for authorization for a Presidential Permit authorizing the constructions, connection, operations and maintenance of pipeline facilities for the importation of crude oil at the United State-Canada border. The new Keystone XL pipeline project would cross the border at Phillips County, MT, and extend to a point on the existing Keystone pipeline system at Steel City, NE. The changes to the route involve only the segment through Nebraska.

As with other Presidential Permit applications, the State Department asserted its need to determine if granting a permit for the proposed pipeline is in the national interest. Consistent with E.O. 13337, that determination will involve the consideration of factors, including energy security, health, environmental, cultural, economic, and foreign policy concerns. To do so, the Department stated that it would select an independent third-party contractor to assist in that process, including a review of the existing EIS from the previous Keystone XL pipeline permit process, as well as identifying and assisting with new analysis.[45] Also, the State Department reaffirmed that it would cooperate with the state of Nebraska, as well as other relevant state and federal agencies, in the review of the application. Finally, the department acknowledged that Nebraska officials expect their review of the new proposed route to take six to nine months. As a result, the State Department estimates that it would complete the review process in the first quarter of 2013.

State Siting and Additional Environmental Requirements

As stated above, the federal government does not currently exercise siting authority over oil pipelines. Instead, siting for the Keystone XL pipeline must comply with any applicable state law—which can vary from state to state. South Dakota, for example, required TransCanada to apply for a permit for the Keystone XL pipeline from the state public utility commission, which issued the permit on April 25, 2010.[46] Montana requires a certificate from the state's Department of Environmental Quality.[47]

At the time of TransCanada's application for a Presidential Permit, Nebraska did not have any permitting requirements that applied specifically to the construction and operation of oil pipelines, although a state statute does include an "eminent domain" provision, which grants eminent domain authority to oil pipeline companies that are unable to obtain the necessary property rights from the relevant property owners.[48] However, due to the controversy surrounding the Keystone XL project, the Nebraska Governor called a special session of its legislature to enact legislation to assert state authority over pipeline siting. Subsequently, the state enacted two laws—one that would affect the siting of the Keystone XL pipeline (see Table 1) and one that outlines procedures for siting any future oil pipeline in Nebraska.[49] The latter will require oil pipeline carrier proposing to construct a major oil pipeline in Nebraska to file an application with the state's Public Service Commission and receive approval before beginning construction. Additionally, the law authorized the Commission to follow certain procedures before deciding whether a proposed oil pipeline would serve the public interest.

Although there are limited federal requirements applicable to oil pipeline *siting*, there are numerous local, state, tribal, and federal requirements applicable to pipeline construction, operation, and maintenance. For example, the August 2011 final EIS for the Keystone XL pipeline identified a list of permits, licenses, approvals and consultation that would be required before the pipeline project could proceed.[50] From that list, following are a few of the requirements that would likely apply to any pipeline project, listed by agency with jurisdiction over that requirement:

- The U.S. Army Corps of Engineers—issuance of a permit for sections of the project that require placement of dredge and fill material in waters of the United States, including wetlands (pursuant to Section 404 of the Clean Water Act), or for pipeline crossings of navigable waters (pursuant to Section 10 of the Rivers and Harbors Act);
- The Environmental Protection Agency—review and issue National Pollutant Discharge Elimination System permits for the discharge of pollutants in state waters (pursuant to Section 402 of the Clean Water Act);
- The Bureau of Land Management—grant temporary use permits for portions of the project that would encroach on federal lands;
- U.S. Fish and Wildlife Service—consider impacts to federally listed endangered species (pursuant to the Endangered Species Act) and provide a Biological Opinion if the project is likely to adversely affect federally listed species.

- Multiple state/county agencies—consult on and/or consider issuance of permits for projects that cross navigable waters or state highways, or involve work potentially affecting state streams, cultural resources, or natural resources.

The time it takes to complete the NEPA process has been a focus of attention for the first Presidential Permit application for the Keystone XL pipeline. However, for past pipeline projects, obtaining all required local, state, tribal, and federal permits, approvals, and licenses may take a similar amount of time. By way of example, for the Alberta Clipper pipeline project, another oil sands pipeline, the Presidential Permit process (i.e., completion of the NEPA process and the national interest determination and issuance of a permit) took approximately two years. Obtaining the necessary permits, approvals, and licenses for construction of the pipeline took an additional two years.

Legislative Efforts to Change Permitting Authority

In light of the State Department's denial of the Keystone XL permit, some in Congress have sought alternative means to support development of the pipeline. As stated in the Introduction, there are a number of legislative proposals to change the federal permitting authority for the pipeline. Key among these is the Surface Transportation Extension Act of 2012, Part II (H.R. 4348), which passed in the House on April 18, 2012. This bill (and H.R. 3548) would transfer the permitting authority over the Keystone XL pipeline project from the State Department to the Federal Energy Regulatory Commission (FERC), requiring the commission to issue a permit for the project within 30 days of enactment. Other proposals, such as H.R. 3811, would directly shift permitting authority to Congress, approving the original permit application filed by TransCanada in 2008 upon enactment.

Changing, or eliminating altogether, the State Department's role in issuing cross-border infrastructure permits may raise questions about the President's executive authority (further discussed in the Appendix A). In response to H.R. 3548, for example, the State Department's key official on Keystone XL testified before Congress:

> The legislation raises serious questions about existing legal authorities, questions the continuing force of much of the federal and all of the state and local environmental and land use management authority over the pipeline, and overrides foreign policy and national security considerations implicated by a cross border permit, which are properly assessed by the State Department.[51]

Such proposals may also raise some administrative and legal challenges for FERC or other federal agencies. A senior FERC official testified that a proposal like H.R. 3548 does not provide enough time for an "adequate" public record, provides no clear authority for enforcing measures required in the EIS, does not articulate a process for authorizing alterations to the pipeline route in Nebraska, and is unclear about permits required from other federal agencies, among other concerns.[52]

Given the State Department's initial permit denial, and opposition from various environmental groups and stakeholders along the pipeline route, legal challenges are a possibility. However, in the event of a challenge based on an environmental issue, the distinction between State Department actions required under NEPA and those required under its authority to issue a Presidential Permit would be relevant. NEPA does not create a private right of action. Instead, judicial challenges to a federal agency action under NEPA are brought pursuant to the Administrative Procedure Act (APA, 5 U.S.C. §§706 et seq.). Presidential actions, however, are not subject to judicial review under the APA.[53] That is, the final agency action reflected in a ROD is subject to judicial review, but the State Department's national interest determination, made under its authority to issue a Presidential Permit, is not. For more information regarding the State Department's authority to grant a Presidential Permit, see the Appendix A.

ARGUMENTS FOR AND AGAINST THE PIPELINE

Proponents of the Keystone XL pipeline, including Canadian agencies and U.S. and Canadian petroleum industry stakeholders, base their arguments supporting the pipeline primarily on increasing the diversity of the U.S. petroleum supply and economic benefits to the United States, especially job creation. Opponents to the pipeline are generally environmental organizations and community groups. Their concerns stem from issues that can be broadly categorized as the pipeline's global or domestic impacts. "Global" impacts stem primarily from concern regarding the lifecycle greenhouse gas (GHG) emissions associated with the development and extraction of Canadian oil sands, compared to conventional oil or renewable fuels. Although the concern regarding GHG emissions is focused primarily on the extraction process, opponents also argue that use of the oil sands crude promotes continued U.S. dependency on fossil fuels. Concern over adverse domestic impacts of the pipeline stem primarily from impacts associated with the pipeline's construction and long-term use on private land—particularly its potential to affect agricultural uses and cattle grazing. Communities along the pipeline route are also concerned about the risk of a potential release of heavy crude and the operators' ability to respond to a release, particularly in remote areas.

Impacts to the Nebraska Sand Hills

In the process of examining factors necessary to determine whether the Presidential Permit for the original Keystone XL pipeline route was in the national interest, the State Department decided that it needed to assess potential alternative pipeline routes that would avoid the Sand Hills region of Nebraska. Unique characteristics of the Sand Hills, including its high concentration of wetlands, extensive areas of very shallow groundwater, and its sensitive ecosystem, were identified as factors that resulted in increasing public concern over the proposed pipeline location. For these reasons, TransCanada announced it would work with the Nebraska Department of Environmental Quality to identify a potential pipeline route that would avoid the Sand Hills.

To understand the concerns associated with potential environmental impacts of the construction and operation of a pipeline that crosses the Sand Hills (also referred to as the Sandhills), an understanding of the unique size and structure of the region is useful. The Sand Hills region is a 19,600 square mile sand dune formation stabilized by native grasslands that cover 95% of its surface. The surface is highly susceptible to wind erosion if the grassland is disturbed.[54] Below its surface lie hundreds of feet of course sand and gravel. Essentially, the porous soil acts like a giant sponge that quickly absorbs precipitation, allowing very little to run off. In some areas, the water table reaches the land surface—a characteristic that creates lakes that dot the region as well as 1.3 million acres of wetlands. The loose, porous soil and sensitivity to wind erosion have been factors contributing to a lack of development on the Sand Hills. As a result, the region contains the most intact natural habitat of the Great Plains of the United States. The porosity of the soil is also relevant because the Sand Hills sits atop the Ogallala Aquifer—one of the largest aquifer systems in the world.[55]

The highly porous soil of the Sand Hills make it a significant recharge zone in the northern region of the Ogallala Aquifer system. That is, the sandy, porous soil of the Sand Hills allows a significant amount of surface water to enter (recharge) the aquifer system. Water from the aquifer also accounts for a significant amount of water use—78% of the region's public water, 83% of irrigation water in Nebraska, and 30% of water used in the United States for irrigation and agriculture.

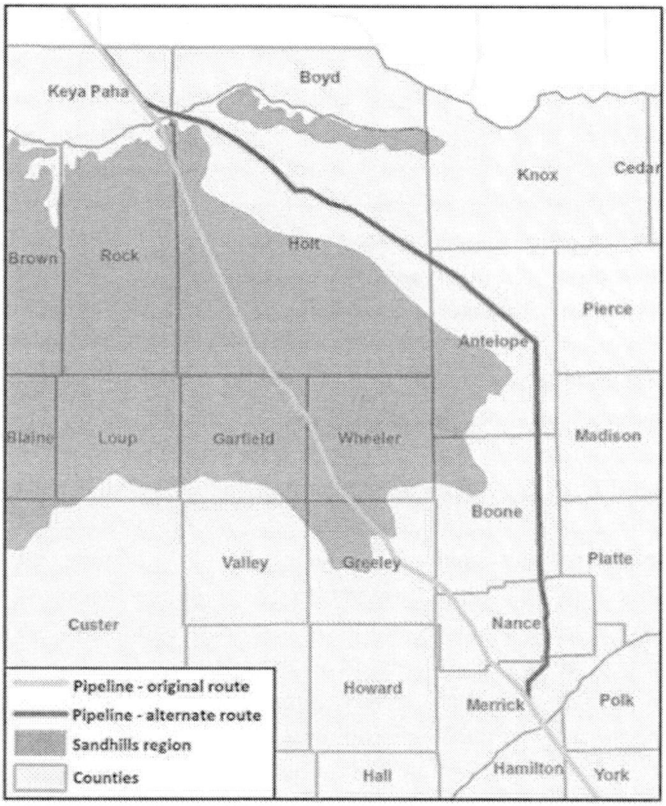

Source: Congressional Research Service, adapted from TransCanada, *TransCanada Keystone XL Pipeline Project, Initial Report Identifying Alternative and Preferred Corridors for Nebraska Reroute*, April 18, 2012, http://www.eenews.net/assets/2012/04/19/document_cw_02.pdf.

Figure 2. Keystone XL Preferred Alternative Route in Nebraska.

Potential impacts to the Ogallala Aquifer and the Sand Hills identified in the final EIS for TransCanada's original permit application included potential groundwater contamination after a release (e.g., a spill or leak from a hole or damaged portion of the pipeline) of crude oil during the construction or operation of the proposed pipeline. Along the preferred route of the originally proposed pipeline configuration, areas in the Sand Hills region were identified as locations where the water table may be close to the surface. The depth to groundwater was less than 10 feet for approximately 65 miles of the preferred pipeline route in Nebraska. Both the soil porosity and the close proximity of groundwater to the surface increase the potential that a release of oil from the pipeline could contaminate groundwater in the region.[56]

The new route alternative TransCanada has proposed for the Nebraska section of the Keystone XL pipeline avoids the Sand Hills (Figure 2). However, the pipeline would still cross part of the Ogallala aquifer.

Impact on U.S. Energy Security

In its Presidential Permit application, TransCanada asserts that constructing the proposed Keystone XL pipeline is in the U.S national interest to maintain adequate crude oil supplies for U.S. refineries. The application argues that the pipeline will allow U.S. refiners to substitute Canadian supply for other foreign crude supply and to obtain direct pipeline access to secure and growing Canadian crude output. In particular, the application asserts that the pipeline would allow the United States to decrease its dependence on foreign crude oil supplies from Mexico and Venezuela, the two largest oil importers into the U.S. Gulf Coast.[57] Consistent with this argument, H.R. 3900 would seek to ensure that any crude oil and bitumen transported by the Keystone XL pipeline, or any resulting refined products, would have to remain in U.S. markets subject to a Presidential waiver allowing foreign export.[58] Depending upon the circumstances, however, such restrictions could raise concerns with respect to international trade agreements, among other considerations.

Energy security arguments have taken on additional weight in light of the recent geopolitical tensions in the Middle East and North Africa. However, it is worth noting that even if Keystone XL is built, prices for the crude oil it carries as well as for domestically produced oil from elsewhere will continue to be affected by international events. The oil market is globally integrated and events in major producer and consumer countries can affect prices everywhere.[59] For example, the disruption of Libyan supply in early 2011 contributed to higher crude oil prices in the United States, even though the United States imported almost no oil from Libya before the unrest broke out.[60]

Canadian Oil Imports in the Overall U.S. Supply Context[61]

Gross U.S. imports of crude oil and petroleum products averaged 11.4 million bpd (Mbpd) in 2011.[62] U.S. oil exports averaged 2.9 Mbpd (almost entirely petroleum products), leaving net imports at 8.4 Mbpd.[63] U.S. net imports have fallen by 4.1 Mbpd or 33% since they peaked in 2005 as a result of lower total oil consumption and higher domestic production. Some of this decline could be mitigated in the near term as oil demand recovers from the recession. However, there is increasing sentiment among forecasters that U.S. oil imports have passed their high water mark already and may remain relatively flat or fall in the foreseeable future.[64]

Among the largest sources of U.S. gross oil imports are Canada (2.7 Mbpd), the Persian Gulf (1.9 Mbpd), Mexico (1.2 Mbpd), and Venezuela (0.9 Mbpd). Imports from the latter two sources have decreased in recent years in part due to lower need for imports described above and in part due to developments in those countries. Mexican production has been falling since 2004 because new oil developments have not been able to offset depletion at Mexico's giant Cantarell field. Imports from Venezuela, another key source of U.S. imports, have also fallen. Venezuelan production never fully recovered after a strike at its national oil company, *Petróleos de Venezuela*, in 2002-2003. Venezuelan production today is nearly 1 Mbpd less than that achieved in 2001. In recent years, Venezuela has also been trying to diversify business away from the United States, for example, by increasing exports to China.[65]

Meanwhile, Canadian production and exports to the United States have increased, primarily due to growing output from the oil sands in western Canada. Energy markets in the United States and Canada are well integrated by pipeline infrastructure; nearly all Canadian energy exports go to the United States.[66] Canadian oil production has increased about 0.2 Mbpd since 2005 and exports to the United States have increased by 0.5 Mbpd (see Figure 3).[67] Some expect Canadian oil production to grow by nearly 2 Mbpd by 2025 due to increased output from the oil sands.[68]

Oil Sands, Keystone XL, and the U.S. Oil Market

Oil sands (also referred to as tar sands) are a mixture of clay, sand, water, and heavy black viscous oil known as bitumen. Oil sands require more processing than conventional crude oil. Oil sands are processed to extract the bitumen, which can then be sent to refineries in one of two forms. Bitumen can be upgraded into "syncrude," a light crude that is suitable for pipeline transport and is relatively easy to refine. Alternatively, bitumen can be blended with lighter hydrocarbons to form a heavy crude (diluted bitumen or "dilbit") that can be transported by pipeline. The bulk of oil sands supply growth is expected to be in the form of the latter.[69]

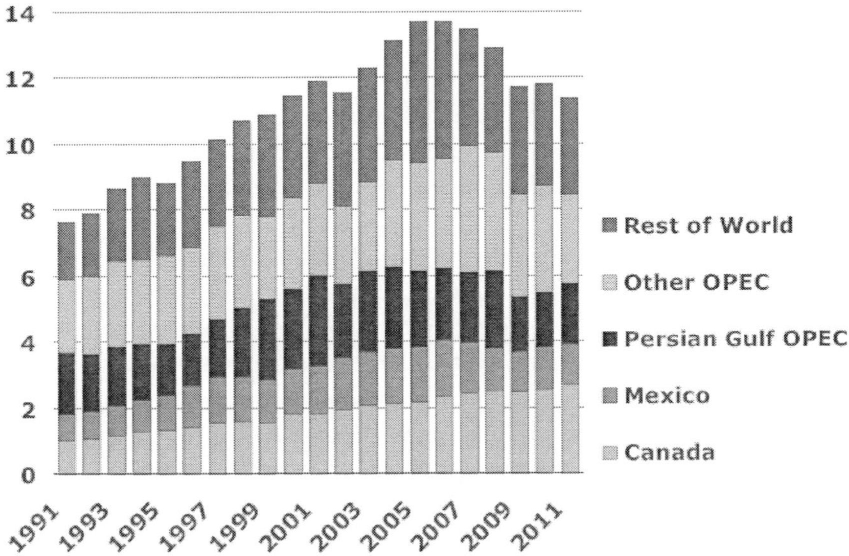

Source: U.S. Energy Information Administration, *Petroleum & Other Liquids: U.S. Imports by Country of Origin*, March 19, 2012. http://www.eia.gov/petroleum/data.cfm#imports.

Figure 3. Gross U.S. Oil Imports by Major Sources; Average annual imports in Mbpd.

Most oil sands imports into the United States currently go to the Midwest, where refineries have been investing in complex refining capacity to process growing volumes of heavy Canadian crude.[70] The U.S. Gulf Coast region already has a large amount of complex refining capacity and is well suited for processing Canadian heavy crude oil. Gulf Coast refiners currently process heavy crudes from Venezuela, Mexico, and elsewhere. Complex refineries in the Gulf Coast may be best equipped to handle a large increase of heavy oil sands crude, though they may still need to adjust processes and make new capital investments in equipment to accommodate particular crudes' characteristics,[71] especially if the new Canadian crudes will be used in large amounts.[72]

There are 15 refineries within Keystone XL's proposed delivery area in Texas that currently process heavy crude oil similar in composition to the oil that Keystone XL pipeline would carry.[73]

Oil production from the oil sands is increasing, as is production from the Bakken and other areas of the U.S. Midwest.[74] Transport options to carry crude from the Midwest to the Gulf Coast are limited. (In the past, crude oil had been shipped up from the Gulf Coast to Midwestern refineries). The resulting abundance of crude oil in the Midwest has driven down crude oil prices in that region relative to Gulf Coast and international crude markets. Midwestern refiners benefit from the lower cost of crude, but it does not translate to substantially lower consumer prices for gasoline or other products in the region. The Midwest still brings in refined products from the Gulf Coast, which keeps refined products prices in line with national and international levels.[75]

Oil sands producers are interested in Keystone XL because it would expand their market reach into the Gulf Coast. The Gulf Coast region holds half of U.S. refining capacity, including a substantial amount of technologically advanced capacity capable of processing heavy sour crudes in large volumes. Reaching a larger market and one with more advanced refining capacity could increase the price these producers receive for their crude. For their part, Gulf Coast refiners are interested in the Keystone XL pipeline because it increases the supply of heavy sour crude in the Gulf region, potentially bringing down their input costs relative to the options they currently have available. Canadian Natural Resources Limited, an oil sands producer, and Valero Energy Corporation, a large U.S. refiner, are among those that contracted for shipping capacity on the Keystone XL pipeline.

With expanded pipeline capacity extending to the U.S. Gulf Coast, Canadian oil sands crude may compete with other heavy crudes such as those from Mexico, Venezuela, and elsewhere.[76] It is difficult to predict precisely how this competition will play out, but it may take place through shifting discounts or premiums on crude oils from various sources.[77] It may be possible for Canadian oil supplies to effectively "push out" waterborne shipments from other countries, although this depends on a wide range of market conditions. Waterborne crudes may more easily go to other destinations than Canadian crudes, though like Canadian crudes they can be tied to specialized refining capacity, as is true for Venezuelan heavy crudes.

There is concern that increased supply of crude to the Gulf Coast may result in larger petroleum product exports rather than contributing to lower domestic fuel cost. Although the United States is a net importer of oil and petroleum products, it does export some petroleum products. U.S. petroleum product exports rose when domestic demand declined in the wake of the recession while foreign demand for certain fuels, such as diesel, remained relatively robust. Issues around potential export of Canadian crude oil carried on Keystone XL or export of products made from that crude oil are addressed in CRS Report R42465, *U.S. Oil Imports and Exports*, by Neelesh Nerurkar.

If Keystone XL secures growing oil sands output for the United States, it could push out seaborne crudes from elsewhere, regardless of where the product is ultimately sold. If the absence of the pipeline encourages Canadian oil sands producers and pipeline companies to find an alternate export route through the Canadian West Coast, Canadian supplies may displace heavy oil supplies in other markets and potentially allow relatively more overseas imports coming into the Gulf Coast. This possibility is discussed further below.

It should be noted that Keystone XL aims to alleviate two potential bottlenecks in the pipeline transportation system: Between Western Canada and the United States, and between the U.S. Midwest and the Gulf Coast. Existing pipelines between Canada and the United States have spare capacity to carry rising Canadian production for the time being. According to some estimates, additional capacity, such as Keystone XL, may not be needed until 2019.[78] The latter bottleneck, between the Midwest and the Gulf Coast, is already at capacity and, as described above, has resulted in a discount for crude oil in the Midwest (though not for petroleum products). The Gulf Coast Pipeline Project, the lower leg of originally proposed Keystone XL pipeline, would address this second bottleneck and help alleviate the discount for Midwestern crudes.

Other Pipeline Projects

Apart from Keystone XL, several other pipeline proposals could help carry growing Canadian crude oil supplies to the U.S. Gulf Coast. On October 16, 2011, Enbridge announced it would purchase ConocoPhillips' share of the Seaway pipeline and reverse its direction to bring crude oil from the Midwest to the Gulf Coast. ConocoPhillips had kept the pipeline running northward to serve its refinery in Ponca City, OK. However, the glut of oil in the Midwest had resulted in the pipeline running at low volumes. Nonetheless, ConocoPhillips had been uninterested in reversing the pipeline. ConocoPhillips, which is spinning off its refining business,[79] sold its share of Seaway to Enbridge. Enbridge and Seaway shareholder Enterprise Products Partners L.P. are reversing the direction of crude oil flows on the Seaway pipeline to enable it to transport oil from Cushing, OK, to the U.S. Gulf Coast. The pipeline is expected to start running southward at an initial capacity of 150,000 bpd starting in May 2012, with capacity expected to increase to 400,000 bpd in 2013. The reversal is expected to reduce the glut of crude oil in the Midwest and reconnect Midwestern crude prices to global prices (driving the U.S. Benchmark West Texas Intermediate crude higher).[80]

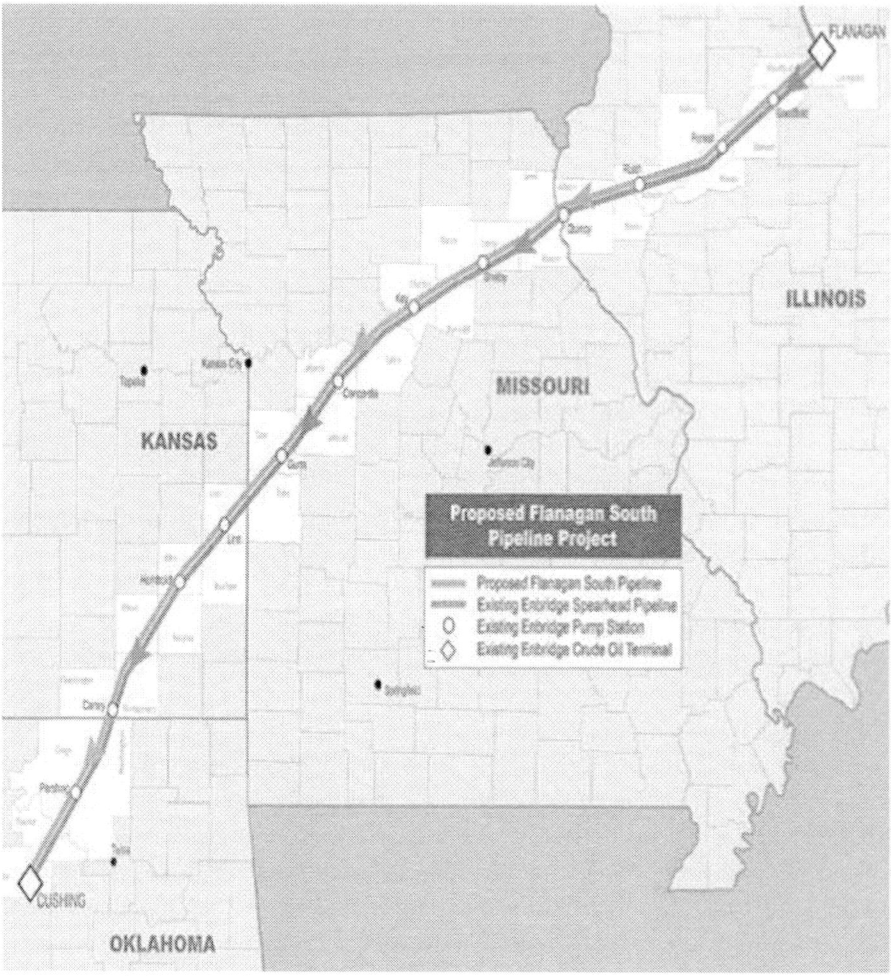

Source: Enbridge, "Flanagan South Project Fact Sheet," April 1, 2012, http://www.enbridge.com/ FlanaganSouthPipeline.aspx.

Figure 4. Proposed Enbridge Flanagan South Pipeline Route.

Prior to the Seaway sale, Enbridge had reported significant commitments for two new pipeline projects: Flanagan South, which would carry oil from Illinois to Oklahoma, and Wrangler, which would carry oil from Oklahoma to Texas.[81] According to Enbridge, the project would duplicate existing routes and would not cross an international border, so it would not require a Presidential Permit. Enbridge already has cross border pipeline capacity connecting Alberta to Illinois. However, according to press reports, Wrangler has been canceled in light of the Seaway purchase and reversal.[82] Enbridge is moving forward with the Flanagan South project, which will have an initial capacity of about 600,000 bpd and run alongside Enbridge's existing Spearhead pipeline (see Figure 4).[83] Like Keystone XL, Flanagan South and a southbound Seaway may facilitate increased flow of Canadian crude to the U.S. Gulf Coast.

Canadian Oil to Alternative Markets

There are proposals to increase the capacity for oil from Alberta to reach the Canadian east and west coast. Currently, nearly all of Canada's oil exports go to the United States, mostly through north-south pipelines. Only one major oil pipeline extends from Alberta to Canada's west coast: the Trans Mountain Pipeline, which is owned by Houston-based Kinder Morgan and has a capacity of 300,000 bpd. Some of the oil from the Trans Mountain Pipeline is loaded onto tankers and shipped from Vancouver. Nearly all of the quantities shipped by sea go to the United States, though a small amount goes to China and other Asian countries.[84] Proposals for additional East and Westbound capacity include:

- Kinder Morgan has plans to expand the Trans Mountain Pipeline to 850,000 bpd by 2017, more than doubling its existing capacity, and expanding west coast shipping facilities.[85] The expansion has received the necessary commitments from parties interested in shipping additional crude volumes. Some shippers are interested in using the additional capacity to export more Canadian crude oil to Asia. Kinder Morgan still needs regulatory approvals from Canadian authorities and is working to gain the support of stake holders.[86] There is some opposition to the project, including from groups concerned about additional tanker traffic near Vancouver and potential oil spill risks.[87]
- Enbridge has proposed a new pipeline: the Northern Gateway project would have a 525,000 bpd capacity to send oil from Edmonton to Kimat, British Columbia.[88] However, Northern Gateway faces

opposition from groups including some First Nations communities and environmental groups.[89]

- Several projects are considering moving oil east rather than to the west coast. According to reports, TransCanada is considering a pipeline project sending oil east from Alberta to Quebec and New Brunswick which could also carry crude bound for export.[90] Enbridge is also interested in expanding eastbound capacity by reversing its Line 9 Pipeline.[91] Some suggest this could potentially lead to oil sands crude traveling east, through Montreal and then through another pipeline to Portland, Maine, from which point it could be exported.[92] As with other pipeline projects, these also face opposition from environmental groups concerned about oil spill risks and/or generally opposed to oil sands development.

These projects reflect anticipated growth of western Canadian oil production and an interest by Canadian oil producers to diversify their available markets beyond U.S. customers, including to reach rapidly growing Asian oil demand. Proposals have received criticism from environmentalists. Because it would require construction of a completely new pipeline, Northern Gateway in particular has been criticized by some environmental and First Nations groups.[93]

Canadian interests assert that Canadian oil sales to Asian markets, where oil demand is growing rapidly, are more likely if greater shipments to the United States are not possible.[94] A study commissioned by the U.S. Department of Energy suggested that:

> if pipeline projects to the BC [British Columbia] coast are built, they are likely to be utilized. This is because of the relatively short marine distances to major northeast Asia markets, future expected growth there in refining capacity and increasing ownership interests by Chinese companies especially in oil sands production. Such increased capacity would alter global crude trade patterns. Western Canadian Sedimentary Basin (WCSB) crudes would be "lost" from the USA, going instead to Asia. There they would displace the world's balancing crude oils, Middle Eastern and African predominantly OPEC grades, which would in turn move to the USA. The net effect would be substantially higher U.S. dependency on crude oils from those sources versus scenarios where capacity to move WCSB crudes to Asia was limited.[95]

Economic Impact of the Pipeline

In addition to supply diversity arguments, some Keystone XL pipeline proponents support the project based on economic benefits associated with expanding U.S. pipeline infrastructure. A recent study by the Energy Policy Research Foundation, for example, concludes that "the Keystone expansion would provide net economic benefits from improved efficiencies in both the transportation and processing of crude oil of $100 million-$600 million annually, in addition to an immediate boost in construction employment."[96] A 2009 report from the Canadian Energy Research Institute (CERI) commissioned by the American Petroleum Institute similarly concludes that

> As investment and production in oil sands ramps up in Canada, the pace of economic activity quickens and demand for US goods and services increase rapidly, resulting in an estimated 343 thousand new US jobs between 2011 and 2015. Demand for U.S. goods and services continues to climb throughout the period, adding an estimated $34 billion to US GDP in 2015, $40.4 billion in 2020, and $42.2 billion in 2025.[97]

These CERI estimates apply to the entire oil sands industry, however, not only the Keystone XL project, and they are derived from a proprietary economic analysis which has not been subject to external review. Some stakeholders point to State Department and other studies reporting much lower anticipated economic benefits.[98] Consequently, it is difficult to determine what specific economic and employment impacts may ultimately be attributable to the Keystone XL pipeline. Nonetheless, given the physical scale of the project, it could be expected to increase employment and investment at least during construction.

Lifecycle Greenhouse Gas Emissions

Oil production from oil sands is controversial because it has significant environmental impacts, including emissions of greenhouse gases during extraction and processing, disturbance of mined land, and impacts on wildlife and water quality.[99] Because bitumen in oil sands cannot be pumped from a conventional well, it must be mined, usually using strip mining or open pit techniques, or the oil can be extracted with underground heating methods.[100] Large amounts of water and natural gas are also required (for heating) during the extraction process.[101] The magnitude of the environmental impacts of oil sands production, in absolute terms and compared to conventional oil production, has been the subject of numerous, and sometimes conflicting, studies and policy papers.[102] Some stakeholders who object to oil sands projects oppose the Keystone XL pipeline because it expands access to new markets for the oil produced by those projects, thereby encouraging what they consider to be further environmentally destructive oil sands development. As discussed earlier, however, if oil sands production can be diverted to other markets (e.g., Asia), preventing the Keystone XL project may not necessarily limit oil sands development.[103]

Some stakeholders object to the Keystone XL pipeline because it would increase U.S. supplies of oil, and thereby perpetuate the nation's dependence on imported fossil fuels and increase carbon emissions from the transportation sector.[104] Acknowledging this concern, in a public forum on October 20, 2010, Secretary of State Clinton reportedly remarked that "we're either going to be dependent on dirty oil from the [Persian] Gulf or dirty oil from Canada ... until we can get our act together as a country and figure out that clean, renewable energy is in both our economic interests and the interests of our planet."[105] Critics of the State Department's draft and supplemental draft EIS assert that the environmental review overlooks the pipeline project's overall impact on greenhouse gas emissions, for example, from the extraction and refining processes. To address those potential emissions, EPA recommended that the final EIS include discussion of mitigation approaches for greenhouse gas emissions from extraction activities that are either currently used or could be employed to help lower lifecycle greenhouse gas emissions.[106] However, others have argued that whether the Keystone XL Pipeline is constructed would have little bearing on greenhouse gas emissions as there are likely to be other export routes available for Canadian oil sands crude, and therefore, the same crude oils would still be transported and refined, albeit in different locations.[107]

Land Use and Oil Spill Impacts

For the pipeline project represented in the August 2011 final EIS, approximately 95% of the land affected by pipeline construction and operation was privately owned, with the remaining 5% almost equally state and federal land. Private land uses were primarily agricultural—farmers and cattle ranchers.

The pipeline's construction and continued operation would involve a 50-foot-wide permanent right-of-way along the length of the pipeline. Keystone agreed to compensate landowners for losses on a case-by-case basis. However, a concern among landowners and communities along the route is the potential for their land or water (used for drinking, irrigation, or recreation) to be contaminated by an accidental release (spill) of oil. That concern is heightened in areas where the pipeline will be located near or would cross water or is in a remote location.

A primary environmental concern of any oil pipeline is the risk of a spill. In estimating the possible impacts of an oil spill, location is generally considered the most important factor— particularly the potential for the spill to reach surface or groundwater. For example, the potential impacts of a spill to water is highlighted in the Keystone XL final EIS, as follows:

> The greatest concern would be a spill in environmentally sensitive areas, such as wetlands, flowing streams and rivers, shallow groundwater areas, areas near water intakes for drinking water or for commercial/industrial uses, and areas with populations of sensitive wildlife or plant species.[108]

A release of oil on land would not necessarily result in surface or groundwater contamination. The potential for a spill to reach water would depend on factors such its proximity to a water source (e.g., on or near a creek or stream or located on land where the groundwater table is close to the surface) and the characteristics of the environment into which the crude oil is released (e.g., porous underlying soils), and the volume of the spill, the duration of the release, and the viscosity and density of the crude oil.

The size of potential spills and the type of oil that would likely be released from the Keystone XL pipeline have been issues of concern to opponents of the project. In its July 16, 2010, comments on the draft EIS for the Keystone XL pipeline, EPA expressed particular concern over the potential adverse impacts to surface and ground water from pipeline leaks or spills. That concern stemmed from two areas—the toxicity of chemical diluents that may be used to allow bitumen to be transported by pipeline and the lack of risk assessment for potential "serious or significant spills," including an evaluation of spill response procedures in the wake of such a spill.

Concerns reflected in EPA's letter were realized 10 days later when the Enbridge Energy Partners' Alberta Pipeline ruptured near Marshall, MI. The resulting spill released dilbit crude into a tributary creek of the Kalamazoo River and traveled approximately 40 miles downstream in the Kalamazoo

River. Initially estimated by Enbridge as a release of approximately 800,000 gallons of crude, EPA subsequently estimated that over 1.1 million gallons were released. The spill resulted in over 220 areas of moderate-to-heavy contamination, including over 200 acres of submerged oil on the river bottom and over 300 solidified oil deposits.[109] Enbridge estimates that cleanup will cost approximately $700 million.

The Enbridge spill highlighted several issues of concern among environmental groups and communities along the pipeline route—in particular, the nature of the dilbit crude likely carried by the Keystone XL pipeline. The dilbit crude in the Enbridge spill had been diluted with benzene and other hazardous constituents. Following the spill, high levels of benzene in the air prompted the issuance of voluntary evacuation of residents in the area. Concern over the presence of similarly toxic constituents, particularly the degree to which the level of toxic constituents may be unknown at the time of a release, has been an ongoing concern among environmental and community groups.

The Enbridge spill was considered a "very large spill" and not necessarily one that would likely occur along the Keystone XL pipeline route. However, in its first year of operation, TransCanada's Keystone pipeline experienced 14 spills. Although mostly minor spills, one spill at the Ludden, ND, pump station resulted in the release of 21,000 gallons of oil. Like the Enbridge release, that release was first reported by local citizens, not as a result of the Keystone's release detection equipment. These incidents have made pipeline opponents concerned that, absent a witness to a spill, a leak in a remote area could potentially go undetected for a long period.

Also as illustrated in the aftermath of the Enbridge spill, cleanup of bitumen crude presents certain challenges. Dilbit is a relatively heavy crude oil mixture compared to other crude oils. In general, heavier oils are more persistent and present greater technical challenges in removal after a spill compared to lighter oils. Almost two years after the Enbridge spill, cleanup efforts continue. Since the spill, public access to 39 miles of the river system was banned to protect public health and safety. The first three-mile segment of river reopened to the public on April 27, 2012. Elements of the cleanup are expected to last until 2015.

Regardless of design, construction, and safety measures, the Keystone XL pipeline will likely have some number of spills over the course of its operating life. The unique oil spill response efforts necessary for dilbit crude make an accurate assessment of potential oil spill risk particularly relevant when addressing concerns expressed by opponents to the Keystone XL pipeline. The need for more conclusive analysis of potential risks associated with the transport of dilbit crude was addressed, in part, in the Pipeline Safety, Regulatory Certainty, and Job Creation Act of 2011 (P.L. 112-90, enacted January 16, 2012). In particular, under Section 16, "Study of transportation of diluted bitumen," the Secretary of Transportation is required to conduct an analysis to determine whether there is any increased risk of a release for pipeline facilities transporting diluted bitumen. In response to that directive, the PHMSA contracted with the National Academy of Sciences to conduct a full and independent study of this topic.

APPENDIX A. PRESIDENTIAL PERMITTING AUTHORITY[110]

The executive branch has exercised permitting authority over the construction and operation of "pipelines, conveyor belts, and similar facilities for the exportation or importation of petroleum, petroleum products" and other products at least since the promulgation of Executive Order 11423 in 1968.[111] Executive Order 13337 amended this authority and the procedures associated with the review, but did not substantially alter the exercise of authority or the delegation to the Secretary of State in E.O. 11423.[112]

However, the source of the executive branch's permitting authority is not entirely clear from the text of these Executive Orders. Generally, powers exercised by the executive branch are authorized by legislation or are inherent presidential powers based in the Constitution. E.O. 11423 makes no mention of any authority, and E.O. 13337 refers only to the "Constitution and the Laws of the United States of America, including Section 301 of title 3, United States Code."[113]

Section 301 simply provides that the President is empowered to delegate authority to the head of any department or agency of the executive branch.

The legitimacy of this permitting authority has been addressed by federal courts. In *Sisseton v. United States Department of State*, the plaintiff Tribes filed suit and asked the court to suspend or revoke the Presidential Permit issued under E.O. 13337 for the TransCanada Keystone Pipeline.[114] The U.S. District Court for the District of South Dakota found that the plaintiffs lacked standing because they would be unable to prove their injury could be redressed by a favorable decision.[115]

The court determined that even if the plaintiff's injury could be redressed, "the President would be free to disregard the court's judgment," as the case concerned the President's "inherent Constitutional authority to conduct foreign policy," as opposed to statutory authority granted to the President by Congress.[116]

The court further found that even if the Tribes had standing, the issuance of the Presidential Permit was a presidential action, not an agency action subject to judicial review under the Administrative Procedure Act (APA).[117] The court stated that the authority to regulate the cross-border pipeline lies with either Congress or the President.[118]

The court found that "Congress has failed to create a federal regulatory scheme for the construction of oil pipelines, and has delegated this authority to the states. Therefore, the President has the sole authority to allow oil pipeline border crossings under his inherent constitutional authority to conduct foreign affairs."[119] The President could delegate his permitting authority to the U.S. Department of State, but delegation did not transform the permit's issuance into an agency action reviewable under the APA.[120]

In *Sierra Club v. Clinton*,[121] the plaintiff Sierra Club challenged the Secretary of State's decision to issue a Presidential Permit authorizing the Alberta Clipper pipeline.

Among the plaintiff's claims was an allegation that issuance of the permit was unconstitutional because the President had no authority to issue the permits referenced in E.O. 13337 (in this case, for the importation of crude oil from Canada via pipeline).[122]

The defendant responded that the authority to issue Presidential Permits for these border-crossing facilities "does not derive from a delegation of congressional authority ... but rather from the President's constitutional authority over foreign affairs and his authority as Commander in Chief."[123] The U.S. District Court for the District of Minnesota agreed, noting that the defendant's assertion regarding the source of the President's authority has been "well recognized" in a series of Attorney General opinions, as well as a 2009 judicial opinion.[124]

The court also noted that these permits had been issued many times before and that "Congress has not attempted to exercise any exclusive authority over the permitting process.

Congress's inaction suggests that Congress has accepted the authority of the President to issue cross-border permits."[125] Based on the historical recognition of the President's authority to issue these permits and Congress's implied approval through inaction, the court found the Presidential Permit requirement for border facilities constitutional.

APPENDIX B. MILESTONES IN THE INITIAL NEPA PROCESS

The NEPA process for TransCanada's 2008 Presidential Permit application for the Keystone XL pipeline project included several significant milestones (summarized in Table 1). These events, and resulting documents, will likely have varying degrees of influence over TransCanada's 2012 permit application.

Draft EIS issued

The State Department released its draft EIS for the proposed Keystone XL Pipeline project for public comment on April 16, 2010.[126] The draft EIS identified TransCanada's "preferred alternative" for the project as well as other alternatives considered. On July 16, 2010, EPA rated the draft EIS "Inadequate."[127] EPA found that potentially significant impacts were not evaluated and that the additional information and analysis needed was of such importance that the draft EIS would need to be formally revised and again made available for public review. Additional criticism of the State Department's implementation of the NEPA process followed an October 21, 2010, statement by Secretary Clinton that, while analysis of the project was not complete and a final decision had not been made, the State Department was "inclined to" approve the project.[128] Critics of the project, including some Members of Congress, stated that the Secretary's statement appeared to prejudge its permit approval for the pipeline proposal as a foregone conclusion.[129]

Supplemental Draft EIS Issued

The State Department issued a supplemental draft EIS on April 15, 2011. In addition to addressing issues associated with EPA's inadequacy rating, the supplemental draft EIS addressed comments received from other agencies and the public. On June 6, 2011, EPA sent a letter to the State Department that rated the supplemental draft EIS as having "Insufficient Information" and having "Environmental Objections" to the proposed action.[130] EPA acknowledged that the State Department had "worked diligently" to develop additional information in response to EPA's comments and the large number of other comments on the draft EIS. However, EPA believed that additional analysis needed to be included in the final EIS to fully respond to its earlier comments.

Among other items, EPA recommended that the State Department should do the following: improve the analysis of the potential oil spill risks, including additional analysis of other reasonable alternatives to the proposed pipeline route; provide additional analysis of potential oil spill impacts, health impacts, and environmental justice concerns to communities along the pipeline route and adjacent refineries; and improve its characterization of lifecycle greenhouse gas emissions associated with Canadian oil sands crude.

In its June 6th letter to the State Department, EPA refers to agreements with the State Department that certain deficiencies identified in the supplemental draft EIS would be addressed in the final EIS. Further, in its conclusion, EPA stated that it would carefully review the final EIS to determine if it fully reflects those agreements and if measures to mitigate adverse environmental impacts are fully evaluated.

Final EIS Issued

On August 26, 2011, the State Department issued the final EIS for the proposed Keystone XL Pipeline. Among other elements of the final EIS, it identified various major pipeline route alternatives and an environmental analyses of potential impacts associated with those alternatives.[131]

In October 2011, 14 Members of Congress wrote to the State Department's Office of Inspector General requesting an investigation of the department's handling of the EIS and National Interest Determination for the Keystone XL project.[132] The request was prompted, in part, by press reports suggesting bias or potential conflicts of interest in the State Department's hiring of an outside contractor to perform the EIS and in its communications with the pipeline's developer, TransCanada.[133] On November 4, the Inspector General's Office (IG) announced that, in response to this request, it was initiating a special review "to determine to what extent the Department and all other parties involved complied with Federal laws and regulations relating to the Keystone XL pipeline permit process."[134] On February 9, 2012, the IG released its findings, reporting that the State Department "did not violate its role as an unbiased oversight agency," among other specific findings generally supportive of the department's Keystone XL permit review process.[135]

Public Review and National Interest Determination

Following the release of the Keystone XL project's final EIS, a review period began to determine if the proposed project was in the national interest. As part of the process for the Keystone XL project, the State Department held public meetings in each of the six states through which the proposed pipeline would pass and in Washington, DC.[136] The meetings were intended to give members of the public additional opportunity to voice their opinions on issues they thought should be taken into account in determining whether granting or denying the Presidential Permit would be in the national interest.[137] During the review period, the State Department received input from state, local, and tribal officials as well as members of the public.

After the public review period, the State Department issued a statement regarding the public comments and its response to those comments.[138] The State Department stated that it received comments on a wide range of issues, including the Keystone XL project's potential impact on jobs, pipeline safety, health concerns, the societal impact of the project, and oil extraction in Canada. Concern regarding the proposed pipeline route through the Sand Hills area of Nebraska was identified as one of the most common issues raised. Comments regarding that pipeline route were consistent with the environmental impacts identified in the final EIS with regard to the unique combination of characteristics of the Sand Hills region.

To understand the concerns associated with potential environmental impacts of the construction and operation of a pipeline that crosses the Sand Hills (also referred to as the Sandhills), an understanding of the unique size and structure of the region is useful. It is a 19,600 square mile sand dune formation stabilized by native grasslands that cover 95% of its surface. The surface is highly susceptible to wind erosion if the grassland is disturbed.[139] Below its surface lie hundreds of feet of coarse sand and gravel. Essentially, the porous soil acts like a giant sponge that quickly absorbs precipitation, allowing very little to run off. In some areas, the water table reaches the land surface—a characteristic that creates lakes that dot the region as well as 1.3 million acres of wetlands.

The Sand Hills sits atop the Ogallala Aquifer—one of the largest aquifer systems in the world.[140] The highly porous soil of the Sand Hills make the area a significant recharge zone in the northern region of the Ogallala Aquifer system. That is, the sandy, porous soil of the Sand Hills allows a significant amount of surface water to enter (recharge) the aquifer system. Water from the aquifer also accounts for a significant amount of water use—78% of the region's public water, 83% of irrigation water in Nebraska, and 30% of water used in the U.S. for irrigation and agriculture.

In the final EIS, the preferred pipeline route through Nebraska would have been located entirely above the Ogallala Aquifer. Potential impacts to the Ogallala Aquifer and the Sand Hills identified in the final EIS include potential groundwater contamination after a release (e.g., a spill or leak from a hole or damaged portion of the pipeline) of crude oil during the construction or operation of the proposed pipeline. Both the soil porosity and the close proximity of groundwater to the surface increase the potential that a release of oil from the pipeline could contaminate groundwater in the region.

During the public review period, the Governor of Nebraska called a special session of the legislature to determine if siting legislation could be crafted and passed for pipeline routing in Nebraska. Facing the prospect of new state pipeline siting regulations applicable to the Sand Hills, together with the concern about the Keystone XL pipeline's specific "preferred" route, the State Department announced that it would require additional information about alternative pipeline routes avoiding the environmentally sensitive Sand Hills area in Nebraska before moving forward with its National Interest Determination.[141] Although the State Department did not decide that environmental issues led to a determination that the proposed project was not in the national interest, environmental issues identified in the final EIS, and further stressed in public comments, led to its decision to delay that determination until it gathered this information.

Although no new decision deadline was established, State Department officials suggested that it would be "reasonable to expect that this process including a public comment period on a supplement to the final EIS consistent with NEPA could be completed as early as the first quarter of 2013."[142] In a prior press interview, President Obama also appeared to suggest that, notwithstanding the delegation of Presidential Permit authority to the State Department, he would be personally involved in the final decision on the Keystone XL Pipeline permit application.[143]

Directive to the President to Approve or Deny the Permit

In the wake of the State Department determination that supplemental analysis was needed, Congress directed the President to make a determination on the Presidential permit application for the Keystone XL pipeline. Specifically, the Temporary Payroll Tax Cut Continuation Act of 2011 (P.L.

112-78), enacted on December 23, 2011, included provisions requiring the Secretary of State to issue a permit for the project within 60 days, unless the President publicly determined the project to be not in the national interest.

Subsequently, the State Department, with the President's consent, announced that it would deny the Keystone XL permit on January 18, 2012. In its announcement the department stated that its decision "was predicated on the fact that [P.L. 122-78] ... passed in December does not provide sufficient time to obtain the information that we think is necessary to assess whether the project, in its current state, is in the national interest."[144] However, the department also stated that its decision did not preclude TransCanada from reapplying for a Presidential Permit in the future, although such a reapplication "will trigger ... a completely new review process."[145]

As a result of that denial, instead of developing a supplemental EIS incorporating analysis applicable to a new pipeline route through Nebraska, a new Presidential permit application process will be required. As a result, a "new" NEPA process will be required. Although much of the analysis and documentation will likely be the same, issuance of a draft and final EIS, and corresponding public and agency comment periods, will be required.

End Notes

[1] This report provides an overview of the Keystone XL project, permit review process, and general policy issues. For more detailed legal analysis, see CRS Report R42124, *Proposed Keystone XL Pipeline: Legal Issues*, by Adam Vann et al. For more analysis of U.S.-Canada energy trade, see CRS Report R41875, *The U.S.-Canada Energy Relationship: Joined at the Well*, by Paul W. Parfomak and Michael Ratner.

[2] U.S. Department of State, "Keystone XL Pipeline Project Review Process: Decision to Seek Additional Information," November 10, 2011, http://www.state.gov/r/pa/prs/ps/2011/11/176964.htm.

[3] U.S. Department of State, "Briefing on the Keystone XL Pipeline," briefing transcript, January 18, 2012, http://www.state.gov/r/pa/prs/ps/2012/01/181492.htm.

[4] TransCanada Corp., "TransCanada Set to Re-Apply for Keystone XL Permit Proceeding with Gulf Coast Project," press release, February 27, 2012.

[5] Ibid.

[6] See the State Department's "New Keystone XL Pipeline Project" webpage at http://www.keystonepipelinexl.state.gov/.

[7] The White House, Office of the Press Secretary, "Statement by the Press Secretary," press release, February 27, 2012, http://www.whitehouse.gov/the-press-office/2012/02/27/statement-press-secretary.

[8] See, for example, Juliet Eilperin, "Democratic Lawmakers Pressure Obama Administration on Both Sides of Keystone Pipeline Issue," *Washington Post*, October 19, 2011; House Energy & Commerce Committee, Subcommittee on Energy and Power, Hearing on The American

Energy Initiative, Discussion Draft of H.R. ____, the North American Made Energy Security Act of 2011, May 23, 2011; U.S. Senator Charles Grassley, Letter to Secretary of State Hillary Rodham Clinton, May 16, 2011; U.S. Senator Max Baucus, Letter to Secretary of State Hillary Rodham Clinton, September 10, 2010; U.S. Representative Henry A. Waxman, Letter to Secretary of State Hillary Rodham Clinton, July 2, 2010.

[9] U.S. Department of State, *Supplemental Draft Environmental Impact Statement for the Keystone XL Oil Pipeline Project*, April 15, 2011. p. 1-4. An initial capacity of 700,000 bpd may be raised to 830,000 bpd by increasing the pumping capacity. The Keystone XL project had applied to the Pipeline and Hazardous Materials Safety Administration to operate at slightly higher pressure than permitted in standard regulations, which would have enabled a 900,000 bpd capacity, but it withdrew its applications for such a Special Permit in August, 2010. The company may reapply for this exemption in the future, however, even after the pipeline is constructed, should it be approved.

[10] TransCanada Corp., "Media Advisory - State of Nebraska to Play Major Role in Defining New Keystone XL Route Away From the Sandhills," press release, November 14, 2011.

[11] TransCanada Keystone Pipeline, L.P., Application of TransCanada Keystone Pipeline L.P. for a Presidential Permit Authorizing the Construction, Operation, and Maintenance of Pipeline Facilities for the Importation of Crude Oil to be Located at the United States-Canada Border, U.S. Dept. of State, September 19, 2008, p. 10, http://www.keystonepipeline-xl.state.gov/clientsite/keystonexl.nsf/presidentialpermitapplication.pdf? OpenFileResource.

[12] "TransCanada Expects $1-Billion Cost Escalation for Keystone XL Pipeline," Canadian Press, February 17, 2011.

[13] Richard M. Pollastro et al., Assessment of Undiscovered Oil Resources in the Devonian-Mississippian Bakken Formation, Williston Basin Province, Montana and North Dakota, 2008, U.S. Geologic Survey, National Assessment of Oil and Gas Fact Sheet (2008–3021), April 2008, p. 1, http://pubs.usgs.gov/fs/2008/3021/pdf/FS08-3021_508.pdf. The Bakken formation also stretches into parts of Manitoba and Saskatchewan, Canada.

[14] Nathan Vanderklippe, "TransCanada to Move U.S. Crude Through Keystone," *The Globe and Mail*, January 26, 2011.

[15] For more on Bakken oil production, see CRS Report R42032, *The Bakken Formation: An Emerging Unconventional Oil Resource*, by Michael Ratner et al.

[16] Jeffrey Jones, "TransCanada Plans U.S. Bakken Pipeline Link," *Reuters*, January 20, 2011.

[17] TransCanada, "TransCanada to Transport U.S. Crude Oil to Market Bakken Open Season a Success," press release, January 11, 2011, http://www.transcanada.com/5631.html.

[18] Vanderklippe, 2011.

[19] Enbridge, "Bakken Pipeline Project—Project Overview," press release, http://www.enbridge.com/ BakkenPipelineProjects/BakkenPipelineProjectUS.aspx.

[20] Lauren Krugel, "TransCanada attracts support for Montana-to-Oklahoma crude pipeline," *The Canadian Press*, January 20, 2011.

[21] Selam Gebrekidan, "Bakken Rail Terminal Ships First Crude Cargo-Lario," *Reuters*, November 9, 2011.

[22] This is in contrast to interstate natural gas pipelines, which, under Section 7(c) (15 USC §717f(c)) of the Natural Gas Act, must obtain a "certificate of public convenience and necessity" from the Federal Energy Regulatory Commission.

[23] See Executive Order 13337, "Issuance of Permits With Respect to Certain Energy-Related Facilities and Land Transportation Crossings on the International Boundaries of the United States," 69 *Federal Register* 25299, May 5, 2004, as amended, and Department of State

Delegation of Authority No. 118-2 of January 26, 2006. The source of Permitting Authority for relevant Executive Orders is discussed further in the Appendix A.

[24] This list was included in the State Department's *Final Environmental Impact Statement for the Keystone Xl Project* under a discussion regarding the Presidential Permit Review Process (p. 1-4). It was noted that this list is not exhaustive, and that the State Department may consider additional factors in its national interest determination process.

[25] In processing Presidential Permit applications, the State Department is also explicitly directed to review the project's compliance with the National Historic Preservation Act (16 U.S.C. §470f), the Endangered Species Act (16 U.S.C. §1531 et seq.), and Executive Order 12898 of February 11, 1994 (59 *Federal Register* 7629), concerning environmental justice. In processing the permit application for the Keystone XL Pipeline project, issues associated with NEPA compliance have drawn the most attention. In large part, that is likely because it is during the NEPA process that compliance with these, as well as any other environmental requirements, would be identified, documented, and demonstrated.

[26] 42 U.S.C. §4332(2)(C).

[27] U.S. Department of State, "Notice of Intent to Prepare an Environmental Impact Statement and to Conduct Scoping Meetings and Notice of Floodplain and Wetland Involvement and to Initiate Consultation under Section 106 of the National Historic Preservation Act for the Proposed TransCanada Keystone XL Pipeline," 74 *Federal Register* 5020, January 28, 2009.

[28] In preparing an EIS associated with a Presidential Permit, NEPA regulations promulgated by both the Council of Environmental Quality (CEQ) and the State Department would apply. CEQ regulations implementing NEPA (under 40 C.F.R. §§1500-1508) apply to all federal agencies. NEPA regulations applicable to State Department actions, which supplement the CEQ regulations, are found at 22 C.F.R. §161.

[29] For more analysis of NEPA requirements, see CRS Report RL33152, *The National Environmental Policy Act (NEPA): Background and Implementation*, by Linda Luther.

[30] Documents submitted for the initial 2008 Presidential Permit application have now been archived by the State Department. Documents related to that original application are available at http://keystonepipeline-xl.state.gov/archive/ index.htm.

[31] 40 C.F.R. §1508.5. Also, Executive Order 13337 directs the Secretary of State to refer an application for a Presidential Permit to other specifically identified federal departments and agencies on whether granting the application would be in the national interest.

[32] For more information, see the U.S. Environmental Protection Agency's "Environmental Impact Statement (EIS) Rating System Criteria" at http://www.epa.gov/compliance/nepa/ comments/ratings.html.

[33] U.S. Department of State press release, "Keystone XL Final Environmental Impact Statement Released; Public Meetings Set," August 26, 2011, http://www.state.gov/r/pa/prs/ps/2011/ 08/171082.htm.

[34] U.S. Department of State, "Keystone XL Pipeline Project Review Process: Decision to Seek Additional Information," Media Note, PRN 2011/1909, Office of the Spokesperson, November 10, 2011.

[35] U.S. Department of State, November 10, 2011, see footnote 34.

[36] The White House, Office of the Press Secretary, "Statement by the President on the State Department's Keystone XL Pipeline Announcement," November 10, 2011.

[37] See TransCanada Corp., Media Advisory, "State of Nebraska to Play Major Role in Defining New Keystone XL Route Away From the Sandhills," November 14, 2011, available at http://www.transcanada.com/5896.html.

[38] For example, see U.S. Department of State, *Record of Decision and National Interest Determination, TransCanada Keystone Pipeline, LP Application for Presidential Permit*, February 25, 2008, http://www.cardnoentrix.com/keystone/ project/SignedROD.pdf.

[39] U.S. Department of State, November 10, 2011, footnote 34.

[40] KETV NewsWatch 7, "Uncut: KETV's Rob McCartney Interviews President Obama," Omaha, NE, November 1, 2011, http://www.ketv.com/video/29652519/detail.html.

[41] U.S. Department of State, Media Note, "Denial of the Keystone XL Pipeline Application," January 18, 2012, available at http://www.state.gov/r/pa/prs/ps/2012/01/181473.htm.

[42] U.S. State Department, Kerri-Ann Jones, Assistant Secretary, Bureau of Oceans and International Environmental and Scientific Affairs, "Briefing on the Keystone XL Pipeline," January 18, 2012, transcript available at http://www.state.gov/r/pa/prs/ps/2012/01/ 181492.htm.

[43] TransCanada Corp. press release, "TransCanada Set to Re-Apply for Keystone XL Permit Proceeding with Gulf Coast Project," February 27, 2012, available at http://www.trans canada.com/5966.html.

[44] See Nebraska DEQ's webpage "Nebraska Keystone XL Pipeline Evaluation: NDEQ's Role in the Pipeline Review Process" https://ecmp.nebraska.gov/deq-seis/.

[45] See the State Department's "New Keystone XL Pipeline Project" webpage at http://www.keystonepipelinexl.state.gov/.

[46] South Dakota Public Utilities Commission, Final Decision and Order; Notice of Entry Before the Public Utilities Commission of the State of South Dakota, In the Matter of the Application by Transcanada Keystone Pipeline, LP for a Permit Under the South Dakota Energy Conversion and Transmission Facilities Act to Construct the Keystone Pipeline Project, HP07-001, http://puc.sd.gov/commission/orders/HydrocarbonPipeline/2008/hp07-001.pdf.

[47] Montana Major Facility Siting Act, Title 75, Chapter 20.

[48] Nebraska Rev. Stat. §57-1101.

[49] See Nebraska Governor Dave Heineman's November 23, 2011 statement "Common Sense Solution," available at http://www.governor.nebraska.gov/columns/2011/11/23_solution. html.

[50] Keystone XL pipeline project final EIS, "Introduction: Section 1.10. Permits, Approvals, and Regulatory Requirements," Table 1.10-1.

[51] Kerri-Ann Jones, Assistant Secretary of State for Oceans and International Environmental and Scientific Affairs, Testimony before the House Energy and Commerce Committee, Subcommittee on Energy and Power Hearing on the North American Energy Access Act, January 25, 2012.

[52] Jeff Wright, Director, Office of Energy Projects, Federal Energy Regulatory Commission, Testimony before the House Energy and Commerce Committee, Subcommittee on Energy and Power Hearing on the North American Energy Access Act, January 25, 2012.

[53] While the APA's definition of "agency" does not specifically exclude or include the president, the Supreme Court has held that exercises of presidential authority are not subject to judicial review because the president is not an agency (*Dalton v. Specter*, 511 U.S. 462, 470 (1994)). The Court has also held that the APA does not apply to the president based on separation of powers principles (*Franklin v. Massachusetts*, 505 U.S. 788, 800-01 (1992)).

[54] For more information, see the Department of the Interior's U.S. Fish and Wildlife Service web page on the Sand Hills at http://www.fws.gov/mountain-prairie/pfw/ne/ne4.htm.

[55] The entire Ogallala Aquifer system stretches across eight states generally from north to south to include South Dakota, Nebraska, Wyoming, Colorado, Kansas, Oklahoma, New Mexico, and Texas and underlies about 174,000 square miles.

[56] Generally, a release of crude oil to land would not necessarily result in groundwater contamination. In addition to the depth from the land surface to groundwater and the characteristics of the environment into which the crude oil is released (e.g., characteristics of the underlying soils), the potential for crude oil to reach groundwater would depend on factors such as the volume of the spill, the duration of the release, and the viscosity and density of the crude oil.

[57] TransCanada Keystone Pipeline, L.P., September 19, 2008, pp. 6-8.

[58] On February 7, 2012, the House Energy and Committee rejected an amendment to H.R. 3548 offered by Representative Edward Markey containing similar export restrictions.

[59] This is the case unless the oil is stranded due to transport bottlenecks. Ironically, the bottleneck for crude oil flowing south from the Midwest to the Gulf Coast—which Keystone XL would help alleviate—helped insulate Midwestern crude oil prices from the impacts of unrest in the Middle East and North Africa. However, as is discussed below, this may have benefited Midwestern refiners but probably did not significantly reduce costs for U.S. consumers.

[60] For more about this, see CRS Report R41683, *Middle East and North Africa Unrest: Implications for Oil and Natural Gas Markets*, by Michael Ratner and Neelesh Nerurkar.

[61] For a primer on the oil market, see CRS Video Brief *Introduction to the Oil Market*, at http://www.crs.gov/analysis/ Pages/WVB00002.aspx.

[62] All data in this section are from the U.S. Energy Information Administration's (EIA's) *Petroleum & Other Liquids* (http://www.eia.gov/petroleum/data.cfm), *International Energy Statistics* (http://tonto.eia.doe.gov/cfapps/ipdbproject/ IEDIndex3.cfm), and the *Short Term Energy Outlook* (http://www.eia.gov/forecasts/steo/).

[63] For context, the United States consumed 18.8 Mbpd in 2011, more than 20% of the world's oil market. Net imports are gross or total imports less total exports. This section will focus on gross imports, though it should be noted that among U.S. petroleum exports about 0.2 Mbpd of petroleum products go to Canada and 0.4 Mbpd to Mexico.

[64] For more analysis, see CRS Report R41765, *U.S. Oil Imports: Context and Considerations*, by Neelesh Nerurkar.

[65] U.S. Energy Information Administration, "Country Analysis Brief: Venezuela," February 2010, http://www.eia.doe.gov/emeu/cabs/Venezuela/Oil.html.

[66] For further analysis of U.S.-Canada energy trade, see CRS Report R41875, *The U.S.-Canada Energy Relationship: Joined at the Well*, by Paul W. Parfomak and Michael Ratner.

[67] As in the United States, Canadian consumption fell due to economic downturn. This allowed the increment in exports to be higher than the increment in production.

[68] Canadian Association of Petroleum Producers (CAPP), *Crude Oil: Forecast, Markets, and Pipelines*, June 2011, p. 2, http://www.capp.ca/forecast/Pages/default.aspx.

[69] CAPP, 2011, p. 7.

[70] CAPP, 2011, p. 13. According to CAPP, refineries adding capacity to process more heavy oil in the Midwest include those in Roxana, IL; Whiting, IN, and Detroit, MI.

[71] Baker Hughes, *Planning Ahead for Effective Canadian Crude Processing*, Baker Petrolite White Paper, 2010, http://www.bakerhughes.com/assets/media/whitepapers/4c2a3c8ffa7e1c3c7400001d/file/28271- canadian_crudeoil_update_whitepaper_06-10.pdf.pdf&fs=1497549.

[72] For a description of which units refineries may need to add (or have added) to be able to process more Canadian oil sands supply, see Praveen Gunaseelan and Christopher Buehler, "Changing US Crude Imports Are Driving Refinery Upgrades," *Oil and Gas Journal*, August 10, 2009.

[73] U.S. Department of State, April 15, 2011. p. 1-4.

[74] See increased U.S. crude oil production in the Midwest under the PADD2 heading at the following source: Energy Information Administration, U.S. Department of Energy, *Crude Oil Production (by PADD)*, Petroleum & Other Liquids, http://www.eia.gov/dnav/pet/pet_crd_crpdn_adc_mbblpd_a.htm.

[75] Adjusted for transport costs and other regional differences.

[76] Center for Energy Economics and Bureau of Economic Geology, *Overview of the Alberta Oil Sands*, University of Texas at Austin, 2006, p. 16, http://www.beg.utexas.edu/energyecon/documents/overview_of_alberta_oil_sands.pdf.

[77] For more about the U.S. refining system, see CRS Report R41478, *The U.S. Oil Refining Industry: Background in Changing Markets and Fuel Policies*, by Anthony Andrews, Robert Pirog, and Molly F. Sherlock.

[78] Testimony of Jim Burkhard, U.S. Congress, Senate Committee on Energy and Natural Resources, *US and Global Energy Outlook for 2012*, 112th Cong., 2nd sess., January 31, 2012.

[79] ConocoPhillips, "ConocoPhillips Pursuing Plan to Separate into Two Stand-Alone, Publicly Traded Companies," press release, July 14, 2011, http://www.conocophillips.com/EN/newsroom/news_releases/2011news/Pages/07-14- 2011.aspx.

[80] Jenny Gross, "NYMEX Oil Gets Boost From Pipeline Reversal," *Wall Street Journal*, April 22, 2012.

[81] Bradley Olson , "Enbridge Pursuing Alternative to Transcanada's Keystone XL," *Bloomberg*, November 9, 2011.

[82] Ben Lefebvre, "Enterprise Products Cancels Wrangler Pipeline," Dow Jones Newswires, November 16, 2011.

[83] Enbridge, "Flanagan South Project Fact Sheet," April 1, 2012, http://www.enbridge.com/FlanaganSouthPipeline.aspx.

[84] According to the Global Trade Atlas, about 0.5% of Canadian crude exports went to China in 2011 (accessed April 25, 2012).

[85] Christopher Smith, "KMEP Advances Trans Moutain Crude Pipeline Expansion," *Oil & Gas Journal*, April 6, 2012.

[86] David Ebner and Justine Hunter, "U.S. Company Plans Billion-Dollar Expansion of Trans Mountain Pipeline," *The Global and Mail*, April 13, 2012.

[87] Jeff Lee, "Vancouver Council, Park Board to Formally Oppose Kinder Morgan Pipeline Expansion," *Vancover Sun*, April 24, 2012.

[88] Enbridge, "Northern Gateway at a Glance," press release, 2011, http://www.northerngateway.ca/project-info/northern-gateway-at-a-glance. The project would also include a pipeline to allow the import of 193,000 bpd of condensate, a light hydrocarbon that can be blended with bitumen to allow pipeline transport.

[89] "Enbridge Pipeline and Tanker Opposition Mounts as Risks Multiply," *Marketwire*, March 13, 2012.

[90] Nathan Vanderklippe and Shawn McCarthy, "TransCanada Looks East as Gateway Pipeline Gets Bogged Down," *The Globae and Mail*, March 22, 2012.

[91] "Enbridge Pipelines Inc. - Line 9 Reversal Phase I Project (OH-005-2011)," (Project Application), National Energy Board (Government of Canada), http://www.neb-one.gc.ca/clf-nsi/rthnb/pplctnsbfrthnb/nbrdgln9phs1/nbrdgln9phs1- eng.html.

[92] Matt Dodge, "Court Decision Affects South Portland-Montreal Pipeline," *Maine Biz*, April 3, 2012.

[93] Derrick Penner, "Opposition to Enbridge Northern Gateway pipeline grows," *Vancouver Sun*, December 2, 2010.

[94] Edward Welsch, "TransCanada: Oil Sands Exports Will Go to Asia if Blocked in U.S.," Dow Jones Newswires, June 30, 2010.

[95] EnSys Energy & Systems, Inc., *Keystone XL Assessment: Final Report*, Prepared for the U.S. Department of Energy, Office of Policy & International Affairs, December 23, 2010, p. 118.

[96] Energy Policy Research Foundation, Inc., The Value of the Canadian Oil Sands (....to the United States): An Assessment of the Keystone Proposal to Expand Oil Sands Shipments to Gulf Coast Refiners, Washington, DC, November 29, 2010, p. 2, http://www.eprinc.org/pdf/ oilsandsvalue.pdf.

[97] Canadian Energy Research Institute, *The Impacts of Canadian Oil Sands Development on the United States' Economy, Final Report*, Calgary, Alberta, October 2009, p. vii.

[98] See, for example, Cornell University Global Labor Institute, *Pipe Dreams? Jobs Gained, Jobs Lost by the Construction of Keystone XL*, September 28, 2011; National Wildlife Federation, "TransCanada Exaggerating Jobs Claims for Keystone XL," November 9, 2010, http://www.dirtyoilsands.org/files/Keystone_XL_Jobs_11-09-10.pdf.

[99] For more analysis of oil sands and their environmental impacts, see CRS Report RL34258, *North American Oil Sands: History of Development, Prospects for the Future*, by Marc Humphries.

[100] U.S. Bureau of Land Management, "About Tar Sands," web page, January 11, 2011, http://ostseis.anl.gov/guide/ tarsands/index.cfm.

[101] Cecilia Jamasmie, "The Challenges and Potential of Canada's Oil Sands," *Mining*, September-October 2010, pp. 7- 8.

[102] For an example of contrasting views, see IHS CERA Inc., *Oil Sands, Greenhouse Gases, and US Oil Supply, Getting the Numbers Right*, 2010; and Natural Resources Defense Council, "Setting the Record Straight: Lifecycle Emissions of Tar Sands," November 2010.

[103] For more analysis of oil sands, including the environmental effects of its extraction, see CRS Report RL34258, *North American Oil Sands: History of Development, Prospects for the Future*, by Marc Humphries.

[104] See, for example: Natural Resources Defense Council, *Tar Sands Invasion: How Dirty and Expensive Oil from Canada Threatens America's New Energy Economy*, May 2010.

[105] See Secretary of State Hillary Clinton's "Remarks on Innovation and American Leadership to the Commonwealth Club," San Francisco, CA, October 15, 2010, available at http://www.state.gov/secretary/rm/2010/10/149542.htm.

[106] See EPA's July 16, 2010, letter to the State Department rating the supplemental EIS for the Keystone XL pipeline project, available at http://yosemite.epa.gov/oeca/webeis.nsf/%28PDFView%29/20100126/$file/20100126.PDF. Discussion of the analysis of GHG emissions is included on pp. 3-4.

[107] EnSys Energy & Systems 2010, p. 116.

[108] U.S. Department of State, *Final Environmental Impact Statement for the Proposed Keystone XL Project*, August 2011, p. ES-9.

[109] For more information see EPA's regarding the response to the Enbridge oil spill at http://www.epa.gov/ enbridgespill/.

[110] For a more expansive treatment of this topic, see CRS Report R42124, *Proposed Keystone XL Pipeline: Legal Issues*, by Adam Vann et al.

[111] *Providing for the performance of certain functions heretofore performed by the President with respect to certain facilities constructed and maintained on the borders of the United States*, 33 *Federal Register* 11741, August 16, 1968.

[112] *Issuance of Permits With Respect to Certain Energy-Related Facilities and Land Transportation Crossings on the International Boundaries of the United States*, 69 *Federal Register* 25299, May 5, 2004.

[113] Ibid.

[114] 659 F. Supp. 2d 1071, 1078 (D. S.D. 2009).

[115] Ibid. at 1078.

[116] Ibid. at 1078, 1078 n.5.

[117] See ibid. at 1080-81.

[118] Ibid. at 1081.

[119] Ibid.

[120] Ibid. at 1082.

[121] 689 F.Supp.2d 1147 (D. Minn. 2010).

[122] Ibid. at 1162.

[123] Ibid.

[124] Ibid. at 1163 (citing 38 U.S. Atty Gen. 162 (1935); 30 U.S. Op. Atty. Gen. 217 (1913); 24 U.S. Op. Atty. Gen. 100; and Natural Resources Defense Council (NRDC) v. U.S. Department of State, 658 F.Supp.2d 105, 109 (D.D.C. 2009)). The court in *NRDC* held that the State Department's issuance of a presidential permit under Executive Order 13337 was not subject to judicial review under the Administrative Procedure Act for abuse of discretion because "the issuance of presidential permits is ultimately a presidential action." 658 F. Supp. 2d at 109, 111-12. The court said that to allow judicial review of such decisions would raise separation of powers concerns. Ibid. at 111.

[125] Ibid.; see also Youngstown Sheet and Tube Co. v. Sawyer, 343 U.S. 579 (1952) (establishing a three-part test for analyzing the validity of presidential actions in relation to constitutional and congressional authority).

[126] Documents submitted for the initial 2008 Presidential Permit application have now been archived by the State Department. Documents related to that original application are available at http://keystonepipeline-xl.state.gov/archive/ index.htm.

[127] U.S. Environmental Protection Agency's July 16, 2010, letter to the U.S. Department of State commenting on the draft EIS for the Keystone XL project is available at http://yosemite.epa.gov/oeca/webeis.nsf/%28PDFView%29/ 20100126/$file/20100126.PDF.

[128] See Secretary of State Hillary Clinton, "Remarks on Innovation and American Leadership to the Commonwealth Club," San Francisco, CA, October 15, 2010, available at http://www.state.gov/secretary/rm/2010/10/149542.htm. The statement by Secretary Clinton was actually made in response to a question about the Alberta Clipper pipeline project which received a Presidential Permit from the State Department in 2009; a State Department spokesman later clarified that the Secretary was referring to the Keystone XL pipeline permit approval.

[129] For example, see the October 21, 2010, letter from Senator Mike Johanns to Secretary Clinton expressing his concern that her statement gave the appearance that approval of the pipeline was a foregone conclusion, http://johanns.senate.gov/public/?a=Files.Serve&File_ id=8b090aa5-76fe-41ca-a674-ae9e37db8d36.

[130] U.S. Environmental Protection Agency's June 6, 2011, letter to the U.S. Department of State commenting on the supplemental draft EIS for the Keystone XL project is available at http://yosemite.epa.gov/oeca/webeis.nsf/ %28PDFView%29/20110125/$file/20110125.PDF?OpenElement.

[131] Environmental analysis associated with pipeline project alternatives is provided in Volumes 1 and 2 of the final EIS.

[132] U.S. Senator Bernard Sanders, et al., Letter to The Honorable Harold W. Geisel, Office of Inspector General, U.S. Department of State, October 26, 2011.

[133] See. for example, Elisabeth Rosenthal and Dan Frosch, "Pipeline Review Is Faced with Question of Conflict," *New York Times*, October 7, 2011.

[134] Harold W. Geisel, United States Department of State, Office of Inspector General, "Information Memo for Deputy Secretary Burns," November 4, 2011, http://sanders.senate.gov/imo/media/doc/ Special%20Review%20Keystone%20XL%20Pipeline%20Nov%2020112.pdf.

[135] Harold W. Geisel, United States Department of State, Office of Inspector General, *Special Review of the Keystone XL Pipeline Permit Process*, AUD/SI-12-28, February 2012.

[136] U.S. Department of State press release, "Keystone XL Final Environmental Impact Statement Released; Public Meetings Set," August 26, 2011, http://www.state.gov/r/pa/prs/ps/2011/08 /171082.htm.

[137] These additional public meetings are not part of the NEPA process. Considering the strong public interest in the pipeline proposal (both opposed and in favor), the public hearings were part of the State Department's national interest determination.

[138] U.S. Department of State, November 10, 2011.

[139] For more information, see the Department of the Interior's U.S. Fish and Wildlife Service web page on the Sand Hills at http://www.fws.gov/mountain-prairie/pfw/ne/ne4.htm.

[140] The entire Ogallala Aquifer system stretches across eight states generally from north to south to include South Dakota, Nebraska, Wyoming, Colorado, Kansas, Oklahoma, New Mexico, and Texas and underlies about 174,000 square miles.

[141] U.S. Department of State, "Keystone XL Pipeline Project Review Process: Decision to Seek Additional Information," Media Note, PRN 2011/1909, Office of the Spokesperson, November 10, 2011.

[142] U.S. Department of State, November 10, 2011.

[143] KETV NewsWatch 7, "Uncut: KETV's Rob McCartney Interviews President Obama," Omaha, NE, November 1, 2011, http://www.ketv.com/video/29652519/detail.html.

[144] U.S. Department of State, January 18, 2012.

[145] Ibid.

INDEX

D

E

F

G

H

I

J

K

L

M

N

O

P

Q

R

S

T